T0229891

The Financial Mathematics of Market Liquidity

From Optimal Execution to Market Making

CHAPMAN & HALL/CRC
Financial Mathematics Series

Aims and scope:
The field of financial mathematics forms an ever-expanding slice of the financial sector. This series aims to capture new developments and summarize what is known over the whole spectrum of this field. It will include a broad range of textbooks, reference works and handbooks that are meant to appeal to both academics and practitioners. The inclusion of numerical code and concrete real-world examples is highly encouraged.

Series Editors

M.A.H. Dempster
Centre for Financial Research
Department of Pure
Mathematics and Statistics
University of Cambridge

Dilip B. Madan
Robert H. Smith School
of Business
University of Maryland

Rama Cont
Department of Mathematics
Imperial College

Published Titles

Proposals for the series should be submitted to one of the series editors above or directly to:
CRC Press, Taylor & Francis Group
3 Park Square, Milton Park
Abingdon, Oxfordshire OX14 4RN
UK

Chapman & Hall/CRC FINANCIAL MATHEMATICS SERIES

The Financial Mathematics of Market Liquidity

From Optimal Execution to Market Making

Olivier Guéant

CRC Press
Taylor & Francis Group
Boca Raton   London   New York

CRC Press is an imprint of the
Taylor & Francis Group, an **informa** business

A CHAPMAN & HALL BOOK

CRC Press
Taylor & Francis Group
6000 Broken Sound Parkway NW, Suite 300
Boca Raton, FL 33487-2742

© 2016 by Taylor & Francis Group, LLC
CRC Press is an imprint of Taylor & Francis Group, an Informa business

No claim to original U.S. Government works

Printed on acid-free paper
Version Date: 20160114

International Standard Book Number-13: 978-1-4987-2547-7 (Hardback)

This book contains information obtained from authentic and highly regarded sources. Reasonable efforts have been made to publish reliable data and information, but the author and publisher cannot assume responsibility for the validity of all materials or the consequences of their use. The authors and publishers have attempted to trace the copyright holders of all material reproduced in this publication and apologize to copyright holders if permission to publish in this form has not been obtained. If any copyright material has not been acknowledged please write and let us know so we may rectify in any future reprint.

Except as permitted under U.S. Copyright Law, no part of this book may be reprinted, reproduced, transmitted, or utilized in any form by any electronic, mechanical, or other means, now known or hereafter invented, including photocopying, microfilming, and recording, or in any information storage or retrieval system, without written permission from the publishers.

For permission to photocopy or use material electronically from this work, please access www.copyright.com (http://www.copyright.com/) or contact the Copyright Clearance Center, Inc. (CCC), 222 Rosewood Drive, Danvers, MA 01923, 978-750-8400. CCC is a not-for-profit organization that provides licenses and registration for a variety of users. For organizations that have been granted a photocopy license by the CCC, a separate system of payment has been arranged.

Trademark Notice: Product or corporate names may be trademarks or registered trademarks, and are used only for identification and explanation without intent to infringe.

Library of Congress Cataloging-in-Publication Data

Names: Guéant, Olivier, author.
Title: The financial mathematics of market liquidity : from optimal execution to market making / Olivier Guéant.
Description: Boca Raton : CRC Press, [2016] | Series: Chapman & Hall/CRC financial mathematics series | Includes bibliographical references and index.
Identifiers: LCCN 2015046425 | ISBN 9781498725477 (alk. paper)
Subjects: LCSH: Finance--Mathematical models. | Liquidity (Economics)--Mathematical models. | Money market--Mathematical models.
Classification: LCC HG106 .G84 2016 | DDC 332.6401/51--dc23
LC record available at http://lccn.loc.gov/2015046425

Visit the Taylor & Francis Web site at
http://www.taylorandfrancis.com

and the CRC Press Web site at
http://www.crcpress.com

To my wife Alix, for her endless support

Contents

Preface

We keep moving forward, opening new doors, and doing new
things, because we're curious and curiosity keeps leading us down
new paths.

— Walt Disney

For a long time, the curriculum of most master's degrees and doctoral programs in Quantitative Finance was mainly about derivatives pricing: equity
derivatives, fixed income derivatives, credit derivatives, etc. Same for the academic and professional literature: instead of dealing with financial markets,
financial mathematics was only dealing with financial products, not to say
sometimes only with abstract payoffs.

The situation has changed over the last ten years, because of at least three
factors: (i) the 2007–2008 financial crisis, (ii) the computerization of execution
strategies and the rise of high-frequency trading, and (iii) the recent (and
ongoing) changes in the regulatory framework.

Beyond the standard topics of Quantitative Finance, some new fields and
new strands of research have emerged. An important one is referred to as
market microstructure. Market microstructure used to be the prerogative of
economists, but it is now also a research concern of the financial mathematics
community. Moreover, mathematicians have widened the scope of this literature. In addition to the topics economists used to cover, the literature on market microstructure now covers new topics such as optimal execution, dynamic
high-frequency market making strategies, order book dynamics modeling, etc.
As a side effect, the word "liquidity" is now more present than ever in the
articles of the Quantitative Finance literature, and not to assume infinite and
immediate liquidity in models, as it used to be the case!

The purpose of this book is twofold: first, introducing the classical tools
of optimal execution and market making, along with their practical use; then,
showing how the tools used in the optimal execution literature can be used
to solve classical and new issues for which taking liquidity into account is
important. In particular, we present cutting-edge research on the pricing of
block trades, the pricing and hedging of options when liquidity matters, and
the management of complex share buy-back contracts.

This book is by far orthogonal to the existing books on market microstructure and high-frequency trading. First, it focuses on specific topics that are rarely, or only briefly, tackled in books dealing with market microstructure. Second, it goes far beyond the existing books in terms of mathematical modeling. Third, it builds bridges between optimal execution and other fields of Quantitative Finance.

Except for Parts II and III, which are related to each other,[1] the different parts of the books can be read independently. As a guide for the reader, below is a brief description of the different parts and chapters.

Part I is made of two chapters with absolutely no mathematics:

- Chapter 1 is a general introduction. We describe how and why research on optimal execution and market making has developed. Then, we put in perspective the main questions addressed in this book.

- Chapter 2 is an introduction to the way financial markets work, with a focus on stocks and bonds. A historical perspective is provided together with a detailed description of the current functioning in Europe and in the United States.[2] If you have never heard the words "bid-ask spread," "limit order," "MiFID," "Reg NMS," "dark pool," "reference price waiver," "tick size," "MTF," or "MD2C platform," you should definitely read this chapter.

Part II is the central part of this book. It deals with optimal liquidation strategies and tactics, from theory to numerics, to applications, with a focus on a framework inspired by the early works of Almgren and Chriss.

- Chapter 3 is a general and modern presentation of the Almgren-Chriss framework. This chapter presents the basic concepts on top of which most execution models are built: execution cost functions, permanent market impact, etc. We go beyond the initial model proposed by Almgren and Chriss, in order to cover all cases encountered in practice. This chapter should be read by anyone interested in one of the topics of Part II or Part III.

- Chapter 4 deals with the different types of execution strategies used by market participants: IS (Implementation Shortfall), POV (Percentage Of Volume), VWAP (Volume-Weighted Average Price), etc. We discuss applications of the Almgren-Chriss framework to all types of orders. We also discuss alternative approaches.

---

[1] One needs to read Chapter 3, in Part II, before reading Part III.

[2] Our main focus is on Europe, but we also deal with the case of the United States in order to highlight the main differences.

- Chapter 5 presents extensions of the Almgren-Chriss framework to cover, for instance, the case of a maximum participation rate to the market, or the case of complex multi-asset portfolios.

- Chapter 6 deals with numerical methods to solve the Hamiltonian systems characterizing optimal execution strategies. A simple and very efficient shooting method is presented in the single-asset case. More complex numerical methods are also presented to approximate the optimal trading curves in the case of multi-asset portfolios.

- Chapter 7 goes beyond the Almgren-Chriss framework. Academic research has been very active over the past five years on optimal execution with all kinds of orders: limit orders, market orders, orders sent to dark pools, etc. In this chapter, we present the approaches we find relevant. In particular, execution tactics and models for the optimal placement of child orders are presented and discussed. We also discuss market impact modeling and estimation.

Part III goes beyond optimal liquidation. The goal is to show that one can use the tools developed in Part II for solving standard and nonstandard pricing problems in which liquidity plays a part.

- Chapter 8 deals with block trade pricing, that is, the pricing of a large block of shares. In this chapter, we introduce the concept of risk-liquidity premium: a premium that should be added or subtracted to the Mark-to-Market price in order to evaluate a large portfolio. We believe that this chapter should be read by anyone who attempts to penalize illiquidity in a quantitative model.

- Chapter 9 tackles option pricing. We show how the Almgren-Chriss framework presented in Chapter 3 can be used to build a model for the pricing and hedging of vanilla options. This new pricing and hedging model turns out to be particularly relevant when the nominal of the option is large, or when the underlying is illiquid (in other words, when liquidity matters). If you still believe that replication (Δ-hedging) is the unique/absolute/universal/ultimate tool for option pricing and hedging, you should read this chapter.

- In Chapter 10, we focus on some specific share buy-back contracts, called Accelerated Share Repurchase (ASR). These contracts are, at the same time, execution contracts and option contracts with both Asian and Bermudan features. We show how the Almgren-Chriss framework presented in Chapter 3 can be used to manage these contracts, and why the classical risk-neutral pricing approach misses part of the picture.

Advanced optimal liquidation models are often very close in spirit to models dealing with market making strategies. Part IV is about quantitative models aimed at designing market making strategies for both the bond market and the stock market.

- Chapter 11 is dedicated to market making models. We present in particular the Guéant–Lehalle–Fernandez-Tapia closed-form formula for the quotes of a market maker in the Avellaneda-Stoikov model. Generalizations of the Avellaneda-Stoikov model are discussed and shown to be suited to dealer-driven or quote-driven markets (such as the corporate bond market for instance). We also discuss models for market making on stock markets.

Furthermore, two appendices are dedicated to the mathematical notions used throughout the book. Appendix A recalls classical concepts of mathematical economics. Appendix B recalls classical tools of convex analysis and optimization, along with central ideas and results of the calculus of variations. The book is (almost)[3] self-contained, accessible to anyone with a minimal background in mathematical analysis, dynamic optimization, and stochastic calculus.

Book audience

This book is mainly intended for researchers and graduate/doctoral students in Quantitative Finance – or more generally in applied mathematics – who wish to discover the newly addressed issues of optimal execution and market making. The new edge of Quantitative Finance presented in this book relies on stochastic calculus, but it also uses tools coming from the calculus of variations, and from deterministic and stochastic optimal control. This book will also be useful for quantitative analysts in the industry, who are more and more asked to go from derivatives pricing issues to new topics, such as the design of execution algorithms or market making strategies. This book is partially related to a course the author has been giving since 2011 to Master and PhD students at Université Paris-Diderot and ENSAE.

---

[3]The concept of viscosity solution is sometimes used in the book, without being addressed in appendices. However, except at some points in Chapter 8, the reader does not need to know anything about viscosity solutions for being able to follow the reasoning of the author.

Acknowledgments

I would like to thank many people.

First, Jean-Michel Lasry, for his friendship, and for the hours we have spent together in front of blackboards. He has always been there to share ideas and answer my questions about mathematics and finance.

Pierre-Louis Lions, who, a long time ago, decided to welcome me in his stimulating scientific environment. His assistant at Collège de France, Véronique Sainz – better known as Champo-Lions – also deserves a special thank you.

Charles-Albert Lehalle, who introduced me, back in 2009, to the wonderful world of market microstructure, optimal execution, and market making. This book would not exist without his great influence. He made it possible for me to contribute to these research fields at a time when they were new for mathematicians. It was a pleasure to collaborate with him when we wrote research papers together, and it is always a pleasure to share insights with him.

Yves Achdou, Guy Barles, and Bruno Bouchard, because they have answered many of my (often stupid) questions about mathematics over the last five years.

Jean-Michel Beacco, who uses every endeavor to connect together practitioners and academics. His daily work is of great importance for the French community of Quantitative Finance.

Nicolas Grandchamp des Raux, Global Head of Equity Derivatives Quants at HSBC, and his team – especially Christopher Ulph and Quentin Amelot. During three years, within the framework of the Research Initiative "Modélisation des marchés actions et dérivés" – a research partnership between HSBC France and Collège de France, under the aegis of the Europlace Institute of Finance – my research on optimal execution (beyond optimal liquidation) has been intimately linked to our weekly discussions.

Laurent Carlier, Deputy Head of Fixed Income Quantitative Research at BNP Paribas, and his team – especially Andrei Serjantov – for the time we have spent discussing market making issues.

Jiang Pu, my PhD student, both for his scientific contribution and for his proofreading.

I cannot cite all of them but many of my academic colleagues in Quantitative Finance, in France and outside of France, deserve a warm thank you. In particular, a special thank you goes to Rama Cont for his support; without him, this book would not exist!

All the people who participated in the proofreading of some chapters of this book. Be they all thanked for their remarks and their friendship.

Eventually, I would like to express my gratitude to my beloved wife Alix for her support, her relentless proofreading, and her patience while I was writing this book.

About the author

Olivier Guéant is Professor of Quantitative Finance at École Nationale de la Statistique et de l'Administration Économique (ENSAE), where he teaches many aspects of financial mathematics (from classical asset pricing to advanced option pricing theory, to new topics about execution, market making, and high-frequency trading). Before joining ENSAE, Olivier was Associate Professor of Applied Mathematics at Université Paris Diderot, where he taught applied mathematics and financial mathematics to both undergraduate and graduate students. He joined Université Paris Diderot after finishing his PhD on mean field games, under the supervision of Pierre-Louis Lions. He progressively moved to Quantitative Finance through the publication of research papers on optimal execution and market making.

Olivier is also a renowned scientific and strategy consultant, who has taken on projects for many hedge funds, brokerage companies, and investment banks, including Crédit Agricole, Kepler Cheuvreux, BNP Paribas, and HSBC.

Olivier is a former student of École Normale Supérieure (rue d'Ulm). He also graduated from ENSAE. In addition to his PhD in Applied Mathematics from Université Paris Dauphine – for which he received, in 2010, from the Chancellerie des Universités de Paris, the Rosemont-Demassieux prize of the best PhD in Science – he holds a master's degree in economics from Paris School of Economics. He was also a "special student" and a "teaching fellow" at Harvard University during his doctoral studies.

His main current research interests include optimal execution, market making, and the use of big data methods in finance.

List of Figures

List of Tables

Symbol Description

$(\Omega, \mathcal{F}, \mathbb{P})$ Probability space, assumed to be large enough.

$(\mathcal{F}_t)_{t \in \mathbb{R}_+}$ Filtration in continuous time. Processes are assumed to be adapted to this filtration.

$\mathbb{E}$ Expectation (with respect to the probability measure $\mathbb{P}$).

$\mathbb{E}^{\mathbb{Q}}$ Expectation with respect to another probability measure $\mathbb{Q}$.

$\mathbb{V}$ Variance (with respect to the probability measure $\mathbb{P}$).

$L^{\infty}(\Omega)$ The set of almost surely (a.s.) bounded random variables.

$\mathbb{H}^0(\mathbb{R}, (\mathcal{F}_t)_t)$ The set of real-valued progressively measurable processes.

$(W_t)_t$ Brownian motion.

$(S_t)_t$ Price process.

$(X_t)_t$ Cash account process.

$(q_t)_t$ Process for the number of shares in a portfolio.

$(V_t)_t$ Market volume process.

σ Arithmetic volatility parameter.

Σ Variance-covariance matrix of price increments.

γ Absolute risk aversion parameter.

$L(\cdot)$ Execution cost function.

η, ϕ, ψ Notations used for execution cost functions of the form $L(\rho) = \eta|\rho|^{1+\phi} + \psi|\rho|$.

$H(\cdot)$ Legendre-Fenchel transform of an execution cost function L.

$\rho_{\max}$ Maximum participation rate to the market.

HJB   Hamilton-Jacobi-Bellman.

CARA   Constant absolute risk aversion.

DARA   Decreasing absolute risk aversion.

IARA   Increasing absolute risk aversion.

ODE   Ordinary differential equation.

PDE   Partial differential equation.

MtM   Mark-to-Market.

$C^k(U)$ The set of functions of class C^k on the open set U.

$W^{1,1}(U)$ The set of real-valued absolutely continuous functions on the open set U.

$W^{1,1}(U, \mathbb{R}^n)$ The set of $\mathbb{R}^n$-valued absolutely continuous functions on the open set U.

$L^p(U)$ The set of L^p functions ($p \geq 1$) on the open set U.

$L^{\infty}(U)$ The set of bounded functions on the open set U.

$\partial_\xi f$ Partial derivative of a function f, with respect to a variable ξ.

$\partial^- f$ Subdifferential of a function f.

$x \cdot y$ The inner product of x and y in a Euclidian space.

Part I

Introduction

Chapter 1

General introduction

> The difficulty lies, not in the new ideas, but in escaping from the
> old ones, which ramify, for those brought up as most of us have
> been, into every corner of our minds.

— John Maynard Keynes

Quantitative Finance (also referred to as Financial Mathematics, or Mathematical Finance) is a young science at the frontier between probability theory, economics, and computer science. Despite its short history, Quantitative Finance has already had its Nobel Prize laureates.[1] Furthermore, it has had a major influence on the financial world, far beyond the influence one could expect from a set of quantitative tools.

Since the 2007–2008 crisis, Quantitative Finance has changed a lot. In addition to the classical topics of derivatives pricing, portfolio management, and risk management, a swath of new subfields has emerged, and a new generation of researchers is passionate about systemic risk, market impact modeling, counterparty risk, high-frequency trading, optimal execution, etc.

In this short chapter we provide a brief overview of Quantitative Finance. Then, we situate the topics tackled in this book – optimal execution, market making, etc. – within the current research strands of Quantitative Finance.

1.1 A brief history of Quantitative Finance

1.1.1 From Bachelier to Black, Scholes, and Merton

When exactly did Quantitative Finance emerge as a scientific field? As for any science, there is no official birth certificate, and we could set the starting

---

[1]Robert C. Merton and Myron Scholes have received the Sveriges Riksbank Prize in Economic Sciences in Memory of Alfred Nobel.

point at different dates. One could indeed go back to Bernoulli,[2] Pascal, Fermat, Fibonacci, or even earlier to Ancient Greece, and find some mathematical developments dealing with what would be called today financial instruments, or simply about gambling. That said, it is common to say today that the real father of (modern) Quantitative Finance is a French man called Bachelier.

Certainly inspired by his work at the Paris Bourse, Bachelier defended in 1900 his doctoral thesis entitled "Théorie de la spéculation" under the supervision of Henri Poincaré. In his thesis (see [14]), he developed for the first time a theory of option pricing, using processes very close to what was called later the Wiener process. However, when Bachelier died in 1946, after a career full of pitfalls, his mathematical research applied to finance was not really famous, and certainly unknown to almost all economists.

It is only in the mid-1950s that the work of Bachelier really started to be read by economists. The story is that Savage, the famous statistician, who knew the work of Bachelier and thought that the mathematical tools developed therein could be useful in economics, sent a few postcards to economists to invite them to read Bachelier's work. Samuelson (who has introduced mathematical tools almost everywhere in economics) received one of these postcards, and read the PhD thesis of Bachelier. Although the assumptions of the Bachelier model were questionable (no discounting, prices that could become negative, etc.), the mathematical tools were there.

In the 1960s, research was carried out to price warrants. Samuelson naturally participated in this scientific adventure, along with other economists such as Sprenkle and Boness. They all proposed formulas for the price of a warrant. These formulas were really close in their form to the eventual formula of Black and Scholes, but the major methodological breakthrough was not there. Samuelson and his contemporaries were using the ideas of their time, using expected values (under $\mathbb{P}$, as we could say today), and did not figure out that options could be dynamically replicated.

In fact, the story leading to the Black and Scholes formula is the following.

As Black recounts it in [23], he was working at Arthur D. Little in 1965 and studied the CAPM (Capital Asset Pricing Model).[3] At the end of the 1960s, Black got interested in the pricing of warrants. He naturally used the ideas of the CAPM to figure out a pricing formula. By writing the value of a warrant as a function of time and price (of the underlying), he ended up with a partial differential equation (PDE), but not with the Black-Scholes formula.

---

[2]See the famous St. Petersburg paradox.
[3]The CAPM was co-invented in 1961 by Treynor, who was another employee of Arthur D. Little.

Scholes arrived at MIT at the same period, after finishing his PhD at the University of Chicago, and he got in touch with Black in Boston. They started working together on the option pricing problem. After a few months of tinkering, they found out that the formula obtained by Sprenkle in [163], in the specific case where the risk-free rate is used both for discounting and as the drift in the stock price dynamics, was the solution of the PDE found earlier on by Black: the Black-Scholes formula was born.

Merton was a PhD student of Samuelson in 1970 and he discussed a lot with Black and Scholes, while they were writing a paper to publish their findings.[4] Merton was interested in option valuation and he noticed that a replication portfolio could be built in continuous time, hence proving that the result of Black and Scholes was in fact completely independent from the CAPM. Black and Scholes got the right formula, but Merton introduced replication into the picture.

The fact that the risk associated with an option could be completely hedged away by using a dynamical trading strategy was a major discovery. The subsequent fact that options (or, more generally, complex derivatives products) could be priced by considering the cost of the replication portfolio constituted a methodological breakthrough: pricing and hedging were in fact the two sides of the same coin.

1.1.2 A new paradigm and its consequences

Black and Scholes' paper was published in 1973. It turned up at the right time, when financial agents very much needed new financial products and new tools to manage risk.

Two years before, in 1971, US President Nixon announced the suspension[5] of the dollar's convertibility into gold, triggering the collapse of the Bretton Woods system. In 1973, despite new attempts to go back to fixed exchange rates, all major currencies were floating. The same year, in 1973, the CBOE (Chicago Board Options Exchange) was founded, and was the first marketplace for trading listed options.

Although it took a few years for the new theoretical ideas to be used on trading floors, the world was ready to use a formula that made it possible to manage option books in a very simple manner.

---

[4]The story goes that the paper was initially rejected by the Journal of Political Economy, but Miller and Fama, from the University of Chicago, wrote a letter to the editors to insist on the importance of the findings, and the paper [26] was finally published in 1973, after revisions.

[5]The suspension was initially supposed to be temporary.

Beyond the Black and Scholes formula, which is in fact the tip of a theoretical iceberg, the theory of derivatives pricing based on replication is the second major historical breakthrough in risk management. The first one was mutualization (based on the law of large numbers, and on the central limit theorem), and it led to modern insurance companies. With derivatives pricing, exposure to future states of the world could be traded on markets, and large books of options could be managed, through the dynamical hedging of the residual risk of the book (at least in theory).[6]

The theory of asset pricing based on replication enabled financial intermediaries to propose more and more complex financial contracts to individuals, firms, and institutions. These contracts give firms and institutions a way to hedge their foreign exchange exposure or protect themselves against a price increase in a strategic commodity. They also make it possible for savers to hedge or diversify their risk, and to easily benefit from the difference between their views on the future and those associated with a risk-neutral probability.

In short, the findings of Black, Scholes, and Merton have changed the way people think about financial risk, and it is one of the reasons why derivatives pricing has fascinated several generations of students, lecturers, and researchers in Financial Economics and Quantitative Finance.

1.1.3   The long journey towards mathematicians

Whether Quantitative Finance was born with Black and Scholes, with Bachelier, or beforehand, is a matter of debate, and it does not really matter.

At first, what we call today Quantitative Finance or Mathematical Finance was not separated from classical Financial Economics. Black, Scholes, and Merton were using complex mathematical tools (in particular Ito Calculus), but they were economists (Merton was working under the supervision of Samuelson, and Scholes had a position at MIT in Economics). Black also started his research by using the CAPM, which was introduced independently in 1961 by Treynor, Sharpe, Lintner, and Mossin – the CAPM being itself built upon the Markowitz's portfolio selection model, published in 1952.

The work of Black, Scholes, and Merton was a major breakthrough in Financial Economics, both conceptually and because financial economists of that time were more interested in studying equilibria, hence the initial use of the CAPM by Black.

---

[6] In both cases, it is noteworthy that the value creation is hardly accounted for in national statistics such as the Gross Domestic Product (GDP). The value created is indeed in utility terms rather than in monetary terms.

The road towards Quantitative Finance as a field involving the community of applied mathematicians, and in particular probabilists, is in fact not straightforward. An important step forward is certainly the papers of Harrison and Kreps [99], and Harrison and Pliska [100], in 1979 and 1981 respectively: the authors pointed out the link between the absence of arbitrage opportunity and martingales,[7] but academic papers written by mathematicians only really turned up ten years later.

After Black and Scholes, and before the 1990s, a lot of important models were built by economists. The Cox-Ross-Rubinstein tree-based model [55] made it possible to explain replication and to price options in a very simple manner without relying on Ito Calculus. It had a great role in the industry. In the fixed income area, the Vasicek model [170] was published in 1977 and the Cox-Ingersoll-Ross model [54] in 1985. Outside pure Quantitative Finance, the research in Financial Economics was focused on frictions, incomplete markets, incomplete information, market microstructure (the initial Kyle's model [121] dates back to 1985), etc.

At the very end of the 1980s, mathematicians started to build up Financial Mathematics or Quantitative Finance as an applicative field of probability theory, and then of more classical applied mathematics involving PDEs, optimization and optimal control. The trip from mathematics (with Bachelier) to economists (with Samuelson and then Black-Scholes-Merton) was as simple as sending a postcard, but the return trip was less easy. Some mathematicians started being employed by banks, or getting in touch with the industry as academic consultants, and practical issues progressively turned into theoretical questions raising academic interest.

1.1.4 Quantitative Finance by mathematicians

From the beginning of the 1990s, and until the 2007–2008 crisis, Quantitative Finance has been a great and effervescent field, involving the initial participants (Black's name is everywhere in Quantitative Finance, from fixed income [24] to asset management with the seminal Black-Litterman model [25] that brings together the CAPM and Markowitz's ideas), other economists, and an increasing number of mathematicians.

The focus, on the equity derivatives side, was on improvements of the Black and Scholes model to account for the volatility surface and its dynamics. The local volatility models of Dupire, and Derman and Kani, constituted a major progress in the industry – see [62]. Numerous stochastic volatility models (for instance the seminal Heston model – see [104]) have also been proposed in the

---

[7]They wrote the first version of the fundamental theorem on asset/arbitrage pricing.

literature since the early 1990s. Local stochastic volatility models have been developed later in the 2000s and more recently. Other models have been developed for super-replication in incomplete markets, for instance to take account of transaction costs, or to replace hedging by robust super-hedging when the value of parameters (often the volatility) is uncertain – see the discussion in Chapter 9.

As far as fixed income is concerned, the 1990s have also been a decade of major advances with the new approach proposed by Heath, Jarrow, and Morton [101, 102], and then through the use of the (BGM) Libor market model [33]. Later on, the SABR model [96] had a great success because of the asymptotic formula derived within this model.

At the turn of the millennium, Quantitative Finance was used all over the world to manage huge portfolios of derivatives: equity derivatives, foreign exchange derivatives, fixed income derivatives, but also credit derivatives.

However, credit derivatives books could not simply be managed as equity derivatives books or fixed income derivatives books. In particular, the famous copula model of Li [131], often used in practice with the calibrated "base correlations" to price and hedge Collateralized Debt Obligations (CDOs), should not have been used so blindly. The 2007–2008 crisis has highlighted the danger of the risk-neutral pricing/hedging models when used in highly incomplete markets. It has also highlighted the importance of model risk: when practitioners believe, without any evidence, that most of the risk is captured by their models, they already have one foot in the grave.

1.1.5 Quantitative Finance today

After the subprime crisis, quantitative analysts and mathematicians involved in Quantitative Finance were often lambasted for having used or built models capturing only part of the risk. They were certainly not at the origin of the crisis, but one cannot say they were only scapegoats. Many academic papers were published that addressed the interesting question of credit derivatives pricing and hedging without enough warning about the limited applicability of models. Furthermore, practitioners often did not take the time to step back and analyze the caveats of the models they intended to use, especially when they were proposed in academic papers bearing the signature of famous academics, or when a similar modeling approach was used by their competitors.

Clearly, the financial mathematics community as a whole has played its part in the catastrophe. During and just after the crisis, it was deeply shaken, and there was no point in continuing conducting the same kind of research:

building new credit derivatives pricing models was simply nonsense, and going on marginally improving the existing models in other areas did not appear as a priority. In fact, new research strands very quickly emerged after the crisis.

Because of new practices in the financial industry, an important research field emerged which deals with collateral concerns and the inclusion of counterparty risk in models. Many of the researchers involved in credit derivatives pricing before the crisis work today in this area.

Because the turmoil on the subprime market led to the bankruptcy of some of the largest financial institutions, and to a systemic crisis, systemic risk has also become a major concern of the academic research in Quantitative Finance (and in Financial Economics). In spite of the progress made in risk management, systemic risk was indeed only rarely addressed in research papers. New models have been built to tackle risk in networks, to model contagions, and to better understand the role of clearing houses.

As far as pricing and hedging models are concerned, there is a new interest in transaction costs, in robust (super)-hedging, and more generally in nonlinear approaches – see [95], and the discussion in Chapter 9.

Another strand of research that has emerged over the last ten years is related to the old field of market microstructure – initially studied by economists – and to the emergence of high-frequency trading. This new strand of research is very large today and it involves at the same time specialists of stochastic optimal control, economists, statisticians, and researchers inspired by econophysics.

In this book, we tackle issues related to this large and renewed literature on market microstructure and high-frequency trading, in particular optimal execution and market making. We also consider classical questions of finance, for which we relax the assumption of infinite and immediate liquidity. These topics did not appear with the crisis, but there has been a new interest since the mid-2000s, both due to new regulations – Reg NMS, MiFID, and then recently Basel III – and because many researchers, after the crisis, had their mind available to tackle new issues.

1.2   Optimal execution and market making in the extended market microstructure literature

The topics addressed in this book are recent topics for the community of mathematicians involved in Quantitative Finance. However, they are related to some old topics addressed by the economists who have participated (mainly since the 1980s) to the emergence of an important literature on market microstructure.

1.2.1   The classical literature on market microstructure

In the traditional classification of economists, market microstructure constitutes an area of finance concerned with the price formation process of financial assets, and the influence of the market structure on this process.

In a nutshell, the main goal of the economists involved in this strand of research is to understand the mechanisms by which the willingness to buy and sell assets of the different types of market participants translates into actual transactions, and to understand the resulting price process. They do not focus on the macroeconomic supply and demand for stocks or other assets. Instead, they look into the different black boxes that make the actual transactions possible, and analyze the trading process.

Information is often at the heart of their approach (e.g., the Kyle model [121]). Economists have studied how information is conveyed into prices. They have also studied the impact of asymmetric information on the bid-ask spread and more generally on liquidity. To understand the provision of liquidity and the determinants of bid-ask spreads, economists have modeled the behavior of market makers with static and dynamic models – see for instance the papers by Ho and Stoll [105, 106], Stoll [165], or the paper of Glosten and Milgrom [77].

The translation of information into transactions and prices is obviously related to market impact, and market impact modeling is certainly one of the first topics of the market microstructure literature that has also been studied by academics from other fields than economics, especially econophysicists and statisticians.

The classical market microstructure literature has therefore tackled a wide variety of issues related to the price formation process, and to the frictions arising at the level of exchanges or other market structures.

Another very important research topic[8] in this literature is the influence of the various possible market structures on the price formation process and on market quality. Fifteen years ago, the debate was about continuous-time trading vs. auctions, the various types of orders, the roles of market makers or specialists, etc. Today, the economic research on market microstructure is more focused on the importance of pre-trade and post-trade transparency, the role played by dark pools, the optimal tick size, etc.

In fact, the research agenda is largely set by the changes that have occurred in the market structure over the last ten years. In particular, the fragmentation of the market following Reg NMS in the United States and MiFID in Europe has raised new theoretical questions about the transmission of information, the real interest of competition between venues, the trading rules and trading fee structures of venues, etc. Furthermore, the rise of computerized trading and the important activity of high-frequency traders also raise numerous questions, on the price formation process obviously, but also on the stability of the market (think of the "flash crash" of 2010). On all these topics, economists aim at providing scientific evidence to help decision makers.

1.2.2 An extension of the literature on market microstructure

Statisticians, mathematicians, and econophysicists have recently addressed some theoretical and empirical questions belonging to the classical market microstructure literature, or related to it. New approaches have been proposed for the same problems (although sometimes not with the same angle, because economists are often more focused on equilibrium considerations than the other researchers), and some new problems are today considered part of the (now) multidisciplinary field of market microstructure.

These changes were probably triggered by the automation of trading and the development of computerized execution algorithms, i.e., by the start of the algorithmic trading revolution. New technologies have led to new questions for the modelers, be they mathematicians, (econo)physicists, or even in areas such as operational research or engineering. The first academic papers dealing with optimal execution were those of Bertsimas and Lo [21] in 1998, and Almgren and Chriss [8, 9] in 1999 and 2001.[9] These papers addressed the question of the optimal scheduling to buy or sell a given (large) number of stocks; this was a first step towards the replacement of traders by trading algorithms. This question was not tackled by economists, and was certainly

---

[8]See Chapter 2 for more details on the technical terms.
[9]The book of Grinold and Kahn published in 2000 (see [78]) also dealt with optimal execution.

not part of the market microstructure literature. However, optimal execution problems only make sense when one considers market frictions such as transaction costs, execution costs, and market impact. Therefore, optimal execution could be seen initially as a new area built on top of market impact and execution cost models, and therefore as an extension of classical topics of market microstructure.

At the same time, econophysicists got interested in market impact modeling, and this area ceased to be the prerogative of economists only. However, it is noteworthy that the new entrants were initially more interested in modeling the market impact as the reaction of a physical system to new volume, rather than as the proceed of an (dynamic) equilibrium.

In spite of this new interest that appeared around 2000, the number of papers on market impact modeling and optimal execution only really skyrocketed at the end of the 2000s. Before the 2007–2008 crisis, Quantitative Finance was more dealing with financial products – and in fact with more and more complex payoffs – than with financial markets. The reappearance of the forgotten word "liquidity" during the crisis, the recent changes in the structure of stock markets due to Reg NMS and MiFID, and the rise of high-frequency trading (which is the leading edge of the algorithmic trading revolution mentioned above), brought classical market microstructure questions and the new question of optimal execution onto the tables of mathematicians and statisticians involved in Quantitative Finance.

New modeling frameworks were then proposed to deal with optimal execution. Models with transient market impact were proposed[10] (see Chapter 7) to go beyond the initial framework proposed by Almgren and Chriss (see Chapter 3). Models involving the use of limit orders (see Chapter 7) and dark pools soon followed, along with the associated risk of not being executed.

Numerous optimal execution models are based on parameters such as intraday volatility (or even intraday correlations), transaction and execution cost parameters, probabilities to be executed, and other parameters that need to be estimated using high-frequency data. Therefore, many statisticians and applied probabilists are now involved in the market microstructure literature. They have proposed new methods to filter out the microstructure noise from estimations. They have also introduced advanced techniques based on stochastic algorithms to estimate parameters. They went even further by proposing new models to describe the dynamics of limit order books (for instance with Hawkes processes).

---

[10]In fact, the first model in that direction had already been proposed by economists. Obizhaeva and Wang's article [151] was indeed published in 2013, but it had been a working paper on the Internet since 2005.

Stochastic optimal control techniques are often used in the literature on optimal execution. In particular, numerous stochastic optimal control models have been developed to choose whether to send market orders or limit orders, and where to send these orders.

Instead of only buying or selling stocks as in the case of optimal execution, high-frequency traders are on both sides of the market. Therefore, the modeling approaches and the mathematical tools that are used for tackling optimal execution issues can also be used for dealing with high-frequency trading strategies, although the questions raised are different.

In particular, there is an important literature on high-frequency market making (liquidity provision) strategies. This literature started in 2008 with the paper of Avellaneda and Stoikov [13]. Their model (presented in Chapter 11) boils down to a complex PDE that was solved by Guéant, Lehalle, and Fernandez-Tapia in [88]. Since then, some more realistic models have been proposed, to find the optimal strategies of the financial agents who try to make money out of liquidity provision, especially on the stock market – although models *à la* Avellaneda-Stoikov turn out to be more relevant on quote-driven markets, such as many bond markets (see Chapter 11).

It is noteworthy that the model proposed by Avellaneda and Stoikov is inspired by the model of Ho and Stoll [105] cited above, and published in 1981. This is another piece of evidence that the new research interests of mathematicians involved in Quantitative Finance are deeply related to old concerns and models of the classical economic literature on market microstructure.

1.3 Conclusion

The story of Quantitative Finance as a scientific field is definitely not linear. It has successively involved a mathematician who defended his PhD in 1900, a great number of brilliant economists – two of them having been awarded the Nobel Prize for their achievement – and a large community of applied mathematicians.

Until recently, Quantitative Finance was seen as the use of stochastic calculus and other mathematical tools to price and hedge securities and address risk management issues. Since the 2007–2008 crisis, Quantitative Finance has tackled a larger swath of topics, from counterparty risk, to systemic risk, to optimal execution, and market making.

This book presents some of the approaches proposed in the academic literature for dealing with the execution of large orders, and for designing market making strategies. Furthermore, this book is unique in that it presents how optimal execution models can be used to solve the important question of liquidity pricing (Chapter 8), and to address classical topics of Quantitative Finance such as option pricing and hedging (Chapters 9 and 10).

Before presenting and discussing models, we dedicate the next chapter to the functioning of the two main markets that are addressed in this book: the stock market and the market for bonds. In particular, we present the recent changes in the market structure.

Chapter 2

Organization of markets

> Change is the law of life. And those who look only to the past or present are certain to miss the future.
>
> — J. F. Kennedy

In this chapter, we start by providing an overview of the history of stock exchanges, from the 19th century to the recent period. Then, we put into perspective the current organization of stock markets, with multiple venues and multiple regimes of transparency. Our goal is to introduce the reader to the new organization of stock markets, together with the usual concepts and vocabulary of today's practitioners involved in execution (brokerage companies, cash traders, regulators, etc.). We also present the recent changes in bond markets. Bond markets – which are mainly OTC (Over-The-Counter) markets, very different from most stock markets – are indeed undergoing changes, mainly because of electronification and new regulations. Throughout this chapter, we mainly focus on Europe, but we sometimes consider the case of the United States, to underline the main differences.

2.1 Introduction

Although they have been recurrently criticized and compared to casinos, exchanges have clearly provided over history – and keep providing – several fundamental services to the economy. As any marketplace, an exchange allows the matching of buyers and sellers at a given point in time. However, established exchanges are more than just the marketplaces or the fairs of the Middle Ages. They guarantee that, at a given time, the prices reflect the balance between demand and supply, and that it will continue in the future. Therefore, exchanges have enabled the development of finance: companies and governments knew that they would find investors at the exchanges to finance their projects (through bonds or shares), and savers could expect to be able to sell at a fair price in the future the securities they had purchased – should they want or need to sell.

Exchanges are intimately linked to the history of modern capitalism. Since their global emergence in the 19th century, they have helped to finance a large range of projects, from medium-size businesses to major industrial projects, but also governments. They have also contributed to a better repartition of risk across the world, during the first globalization, and since the start of the second (current) globalization. Exchanges have also played a major role to help people investing confidently their savings, especially in countries with low or declining levels of welfare state.

Exchanges are also strongly linked to the history of technology and telecommunications. The telegraph was utilized to communicate between exchanges in the United States, between exchanges in Europe, and later on to communicate across the Atlantic Ocean with a latency of a few seconds. The telephone then modified the way traders and brokers worked, and therefore the functioning of exchanges. Computers have also changed the way stocks and bonds are traded: most of the traditional stock exchanges have been replaced by electronic systems, in which agents can (virtually) meet to buy and sell securities.[1] More recently, optic fiber and microwaves have made it possible to achieve ultra-low latency in the communication with exchanges, especially for the trading of stocks, where high-frequency traders represent an important part of trading volumes (see [128]), and often act as liquidity providers.

The way stocks and bonds are traded have clearly changed over time, because of the emergence of new technologies. However, technology is not the only factor of change. Regulation is another important component. MiFID (Markets in Financial Instruments Directive) in Europe and Reg NMS (National Market System) in the United States have opened up the market of trading platforms on the equity side. New companies and trading platforms have emerged, so that nowadays the same stock can be sold and purchased on different trading platforms, each of them having its own rules. Bond markets, which mainly remain OTC and dealer-driven, are also undergoing some changes, due to the emergence of electronic platforms. They are also impacted by the new capital requirements applying to investment banks, that deter them from holding large inventories and playing their classical role of dealers.

Furthermore, the upcoming MiFID 2 is going to change the rules once again, with consequences on both bond and stock markets.

In this chapter, we summarize the history of exchanges, from the beginning to the current period. We focus on the organization of the stock and bond markets, along with the current microstructure of trading venues. In addition to being descriptive, we emphasize various economic and financial questions related to microstructure: the quality of the price formation pro-

---

[1] The old trading pits are often exposed in museums.

cess, the different roles played by high-frequency traders, etc. It is noteworthy that we describe the evolution of the organization of markets with a focus on (western) Europe. However, we often mention the differences between Europe and the United States.[2]

Note: The reader interested in the organization of markets will find more details, especially with respect to history, in [97], [98], and [128].

2.2 Stock markets

2.2.1 A brief history of stock exchanges

In this section, we give a short overview of the history of (stock) exchanges. This history has been a common one for stocks and bonds, until recently. Nevertheless, we decided to present it in this section on stocks, not only because stocks happen to be tackled more often than bonds in this book, but also because it permits us to put into perspective the recent changes in the way stocks are traded.

2.2.1.1 From the 19th century to the 1990s

The history of stock exchanges is linked to the history of business. Although the first stock exchanges date back to the Renaissance – exchanges were created in Bruges (15th century), Lyon (16th century), Amsterdam (17th century), etc. – stock exchanges only started to play a fundamental role in the development of western economies in the 19th century, when the capital needs of companies engaging in large-scale and long-term projects (railways, mining, etc.) skyrocketed. The need to rely on individual savers to finance long-term projects, and the need to guarantee investors that securities could quickly and cheaply be transferred, led to the emergence of stock exchanges in major advanced countries. These stock exchanges were often club-like/not-for-profit organizations regulated by strict rules, where membership was given (or sold) to a few selected agents only.

The success of stock exchanges in the 19th century is not only due to the industrial revolutions, and to the global economic development, but also to the trust in the system. This trust was mainly based on the fact that the members – who were the only people to have access to the floor, and therefore to participate in the price formation process[3] – had a lot to lose if they did not abide by the rules.

---

[2]Sometimes, very rarely, we also mention the Asian markets in the book.
[3]Prices were communicated by using the available technological means, and used outside of the stock exchange.

By 1914, most advanced countries had set up their own stock exchange(s). The New York Stock Exchange (NYSE) had existed for a long time, but it only took its current name in 1865. The London Stock Exchange (LSE) was officially founded in 1801. The current Italian stock exchange (Borsa Italiana) is based upon the one that was established in Milan in 1808. In Nordic countries, the Stockholm Stock Exchange was founded in 1863, and the OMX in Helsinki in 1912. Even in the colonial empires, local exchanges were established: Hong Kong in 1891, Jakarta in 1912, etc.

After the four-year disruption caused by World War I, stock exchanges only partially recovered. Furthermore, there was a series of crises in the 1920s and 1930s, in particular the Wall Street Crash in 1929, followed by the progressive end of the Gold Standard and the Great Depression. Investor confidence was shattered, and had to be restored. The main consequence for stock exchanges was the establishment of new rules, and regulatory bodies to enforce them – for instance the Security and Exchange Commission (SEC) created in 1934.

After a new disruption caused by World War II, stock exchanges recovered but the situation had changed. Regulation was indeed imposed or reinforced almost everywhere, and stock exchanges were no longer only in the hands of their members, but also in those of governments willing to regulate markets. In many countries, although not in the United States where the NYSE competed with unregulated markets, stock exchanges were national monopolies, either *de jure* (as in France) or *de facto* (as in the United Kingdom). This situation, with regulated monopolies, was characterized by high fees charged to final investors and almost no evolution. In parallel, the development of a powerful banking system made it possible for firms to be financed without going to the market (especially in Europe).

In the 1970s and 1980s, some powerful institutional investors tried to bypass stock exchanges by trading directly with dealers. They also asked the regulator for changes. In particular, they lobbied for more competition between the members of stock exchanges, in order to reduce transaction costs. However, no real change occurred in Europe, because exchanges were protected, and because members had little reason to abandon their privileges. The only important change in the 1970s was the creation of the NASDAQ in 1971, that aimed at competing with the NYSE. The reaction of the NYSE was the removal of fixed commission charges. In the 1980s, the LSE followed the example of the NYSE, but the degree of competition still remained very low in Great Britain, as elsewhere in Europe.

Real changes occurred in the 1990s, because of the deregulation and privatization waves in many advanced countries, because firms were increasingly relying on financial markets to raise capital (think of the Internet bubble at the end of that decade), and because individuals were increasingly looking for investment opportunities in order to compensate the expected decline of the welfare state, in particular public pension systems. During that decade, the trade-off between the current benefits of a *statu quo* and the opportunities brought by the new economic environment was in favor of the latter. However, most stock exchanges were still organized as clubs, a structure that was not optimal for making decisions, nor for adapting to a new and fast-changing world. Demutualization of most major stock exchanges occurred in the late 1990s, converting the old structures of stock exchanges into standard corporate ones – often into self-listed companies. Demutualization made stock exchanges ready to embrace the new economic environment, and recover the central role they used to have a century before.

2.2.1.2   The influence of technology

In the 1990s, the need for change, in a world impacted by a decade of market freedom, was evident. However, in addition to the economic environment, technology played a major role in triggering the changes.

If the history of stock exchanges is intimately linked to the history of business, it is also intimately linked to the history of telecommunication and technology. The telegraph, invented in 1840, made it possible to communicate over long distances. Later on, at the end of the 19th century, the telephone reduced to seconds the communication latency between the different financial centers.

In 1990, except for one, all major stock exchanges were still not fully electronic. Not surprisingly, the electronification of exchanges started with new exchanges, launched directly in an electronic format, to trade derivative products (futures, options, etc.) in the mid-1980s – see [98]. Electronic trading was then progressively tested on stock exchanges, but the transition was long. Eventually, all national stock exchanges adopted new technologies one after the other, once a few major ones had jumped in.

The electronification of trading had a cost for stock exchanges, but some economies of scale were possible. This is why a wave of mergers occurred, leading to the creation of international stock exchange groups. The most famous one is Euronext, created in 2000 following the mergers of Amsterdam Stock Exchange, Brussels Stock Exchange, and Paris Bourse – the Portuguese stock exchanges were acquired two years later. Later on, in 2007, it merged with NYSE, but turned back into a stand-alone company in 2014. Another example of this consolidation wave is the acquisition of Borsa Italiana by the LSE in 2007. It is noteworthy that consolidation also occurs across all securities, and not only in the equity world.

2.2.1.3   A new competitive landscape: MiFID and Reg NMS

In the old floor-based world, only a few people were able to see the prices and the orders. The electronification of stock exchanges changed this. One can now see the state of the limit order book, or at least part of it (depending on the price one pays to the exchange). This new world of transparency made the stock exchanges look like the ideal markets of economic theory. At the turn of the millennium, buyers and sellers could have detailed, if not full, information about the orders resting in the books,[4] and they knew that the price formation process was working well.

However, despite the potential international competition between stock exchanges, most stocks remained traded on a single venue, often based in the original country of the company.[5] Competition was strong between brokers and between traders, but there was almost no competition between stock exchanges. This situation resulted in high transaction fees.

On both sides of the Atlantic Ocean, regulators decided to boost competition in order to lower transaction costs. In the European Union, MiFID was voted in 2004, and applied effectively from November 2007. MiFID was not only aiming at liberalizing the exchange industry by introducing competition,[6] but also at creating a pan-European market in an attempt to foster market integration. In the United States, Reg NMS (National Market System), was voted in 2005.[7] It set new rules, especially the "order protection rule," that killed the competitive advantage of incumbent exchanges.

Thanks to MiFID, competition within Europe became real. In particular, new operators started proposing their services soon after MiFID. By the end of 2008, it was possible to trade almost all European blue-chips listed on European traditional exchanges, both on the original venue and on alternative ones, which often had (and keep having) the status of MTFs (Multilateral Trading Facilities).[8] Among these MTFs, the most famous ones were operated by Turquoise, BATS, and Chi-X.[9] Turquoise was originally founded by large investment banks, but it was then acquired by the LSE. Chi-X (Europe)

---

[4]This reality is challenged today by the speed of high-frequency trading that makes snapshots of limit order books less and less reliable to make decisions.

[5]There were some cases of multiple listing. See also the ADRs in the United States, although this is different from double-listing.

[6]In many countries, investment firms were required to route the orders of their client towards the national regulated exchanges. This "concentration rule" was abolished by MiFID, wherever it existed, and replaced by the concept of "best execution."

[7]We do not present MiFID or Reg NMS in details in this chapter. Instead, we focus on the main consequences of these regulations.

[8]We adopt here the European vocabulary. In the United States, we talk about ECN (Electronic Communication Network) or ATS (Alternative Trading System) to designate alternative venues – not to confuse with ATS (Average Trade Size).

[9]NASDAQ-OMX also launched an MTF in Europe, but it ceased activity in 2010.

was the first pan-European MTF. It has been acquired by BATS Europe in 2011, another major MTF (BATS being originally from the United States), to form BATS Chi-X Europe. It is important to note that, in spite of mergers and acquisitions, several limit order books are still operated by each company, each venue having specific rules. In other words, mergers and acquisitions have more to do with technology and capitalistic interests than with market fragmentation.

In Europe, following the consolidation wave described above, the big names today are LSE Group, Euronext, Deutsche Börse, and BATS Chi-X Europe. In the United States, although the rules were not the same, Reg NMS led to fierce competition and then consolidation. Today, the big names are NYSE,[10] NASDAQ, and BATS (which recently acquired Direct Edge).

In addition to classical venues organized around limit order books, the above exchange companies, along with some other financial companies, operate another type of trading venue called dark pools (see below) that also represent a significant part of trading flows.

Since 2005, the trading environment has changed dramatically with new exchanges, new rules, and new actors. Although the consolidation wave seems to be over, some new regulations, especially MiFID 2, will soon change the trading environment again.

2.2.2   Description of the trading environment

2.2.2.1   Introduction

In this subsection, we describe the most important features of exchanges (or, more exactly, venues), as of 2015.[11] As always, our focus will be on Europe, although we sometimes speak about other regions to highlight some specificities.

An important consequence of MiFID is that liquidity, for a given stock, is fragmented over several venues. Among these venues, there is the historical exchange of the stock, often called the main or primary venue. It is a Regulated Market (RM) in the MiFID terminology. It is organized around a limit order book (with often the opportunity to post hidden orders – see below) during the continuous trading phase. In addition to this continuous trading phase, there are periods of auctions that involve a lot of agents at the same

---

[10]It is in fact ICE-NYSE, and not NYSE. This proves that consolidation also occurs across asset classes.

[11]Our focus is on stocks, but Exchange-Traded Funds (ETFs) constitute a very close asset class.

time to form a price: opening auctions, closing auctions,[12] and sometimes other auctions during the day, such as volatility interruption auctions – see [128] for more details on auction rules around the world. Furthermore, other venues may be available to trade a given stock. These venues, that are often MTFs, can be organized around a visible limit order book with sometimes the possibility to post hidden orders,[13] or around a trading system that matches buyers and sellers without pre-trade transparency (dark pools).

TABLE 2.1: Fragmentation of the turnover for FTSE 100 stocks (June 8, 2015–June 12, 2015). Source: Fidessa.

Lit pools (without auctions)	41.60%	LSE	52.94%
		BATS Chi-X CXE	22.69%
		Turquoise	15.58%
		BATS Chi-X BXE	7.58%
		Others	1.21%
Auctions	7.42%	LSE	100%
Dark pools	4.88%	BATS Chi-X CXE	22.45%
		Turquoise	17.86%
		BATS Chi-X BXE	16.55%
		UBS MTF	15.66%
		Posit	14.47%
		Instinet BlockMatch	8.57%
		Liquidnet	4.19%
		Others	0.25%
OTC + Systematic Internalizers	46.10%		

The fragmentation of liquidity is really significant on certain stocks. Table 2.1 shows the fragmentation of the turnover on the stocks of the FTSE 100 Index over the period June 8, 2015–June 12, 2015.[14]

Even though there are a lot of venues, the venues of the same type usually follow similar rules. We first describe how limit order books work, and we present the main types of orders usually proposed. Then, we focus on issues related to tick size. We also present the different trading fee structures

---

[12]The closing auctions play an important role in most European exchanges, where they account for 5%–20% of the daily volumes. In other parts of the world, closing auctions may not exist. For instance, in Hong Kong, the closing price is defined as the median of five prices, observed every 15 seconds during the last minute of trading. When there is an auction, it is more and more often the case that the end time of the auction is chosen randomly – in a time window of less than a minute. This is to avoid manipulation on the reference (opening or closing) prices.

[13]Hidden orders inside the book can be regarded as dark liquidity. They are often accounted as lit, because they are eventually executed on lit pools.

[14]Fidessa offers a very nice web interface to measure fragmentation. The definition of each trade category is available at http://fragmentation.fidessa.com/faq/.

applied by the venues organized around a limit order book. In the last paragraphs of this subsection, we finally focus on dark liquidity and the role of high-frequency trading.

2.2.2.2   Limit order books

Principles and basic vocabulary

In a nutshell, an exchange organized around a visible limit order book (LOB) is based on a transparent system that matches the buy and sell orders of market participants on a price/time priority basis.[15] It is a transparent system as all market participants can see, at a given point in time, the limit orders resting in the LOB.[16] It is also very often an anonymous system, because no one can see the identity of the market participants.

Each limit order is characterized by a side (buy or sell), a size (i.e., a number of shares to be bought or sold), and a price. Buy orders resting in the limit order book constitute the bid side of the LOB, while sell orders resting in the limit order book constitute the ask side of the LOB. The highest price of a buy limit order resting in the LOB is called the best bid price, while the lowest price of a sell limit order resting in the LOB is called the best ask price.[17] The average between the best bid price and the best ask price is called the mid-price.

By construction, the best bid price is strictly lower than the best ask price (otherwise, buy and sell orders would have been matched and would have disappeared from the LOB). The difference between the best ask price and the best bid price is called the bid-ask spread.

Table 2.2 shows the state of the limit order book for Siemens on the CXE venue operated by BATS Chi-X Europe. We see that the best bid price is 93.30 € and that the best ask price is 93.32 €.[18]

If an agent is willing to buy a given number of shares, he may decide to send a market order. This market order will be matched with the sell orders resting in the book, on a price/time priority basis. In the case of Table 2.2, if the size of the market order is less than 740, then the market order hits the order(s) resting at 93.32 €, and the agent buys the shares at 93.32 € each. However, if the market order is for 1,000 shares, then the first 740 shares are

---

[15]Other matching algorithms, such as those based on pro-rata rules, may be used on some fixed income markets.

[16]At least, the visible ones.

[17]Sometimes, we simply call bid price and ask price the best bid price and the best ask price.

[18]We do not see in Table 2.2 the number of orders at each price, but only the total volume.

TABLE 2.2: First limits on CXE for Siemens AG – June 15, 2015 (10:16:12).

Buy orders		Sell orders	
Price (€)	Volume	Price (€)	Volume
93.30	340	93.32	740
93.29	387	93.33	1064
93.28	589	93.34	654
93.27	684	93.35	966
93.26	525	93.36	634

paid 93.32 € each, and the following 260 shares are paid 93.33 € each. The agent can also post a limit order at a price strictly lower than the ask price. In that case, no transaction occurs until the order is matched... but this matching may never happen.

The above story is a bit simplistic, compared to what happens in reality, but it sheds light on the basic choice of an agent. He may choose to be a liquidity provider or a liquidity taker. If he provides liquidity, then he has no guarantee that his order will be executed (partially or totally), but the price of the transaction will be the price associated with the limit order (lower than the best ask price in the case of a buy order). If he takes liquidity, then he gets the number of shares he wants (providing there is enough liquidity), and the price of the transaction is at least the best ask price.

In practice, market orders are seldom used. Instead, agents (or algorithms) send limit orders at the best ask price, and these orders turn out to be marketable limit orders if they indeed match resting sell orders.[19] In fact, several types of limit orders are often used instead of standard limit orders for that purpose. Immediate or Cancel (IOC) orders[20] for instance are limit orders matched with the available liquidity resting in the book at the price of the order, but if the order is not entirely filled, the remaining portion is automatically cancelled. Another kind of order is referred to as Fill or Kill (FOK) orders. A FOK order is matched if and only if the order can be entirely filled, otherwise it is automatically cancelled.

Tick size

Another important point that was not highlighted in the above story is that limit orders cannot be posted at any price. Prices have to be multiples

---

[19]An issue, due to high-frequency trading, is that the state of the limit order book often changes between the decision time to send an order and the time when it is received by the trading platform.

[20]We use here the terminology of the LSE. On other platforms, these orders are sometimes called Fill and Kill (FAK) orders.

of a given monetary value, called the tick size (this minimum price increment is 1 cent in the case of the limit order book of Table 2.2, and we see that the bid-ask spread is equal to 2 ticks). The tick size usually depends on the price of the stock (and sometimes also on its liquidity).

Following the tick size war in Europe, the Federation of European Securities Exchanges (FESE) tried to harmonize the different tick size regimes. Although the war has come to an end, different tick size tables are still used by different venues. In particular the FESE proposes four tables, three of which are being used by most of the actors in 2015.

Table 2.3 shows the tick size table no.4 of the FESE, used for instance by Deutsche Börse and Euronext for most stocks. We read in Table 2.3 that, for instance, the minimum price increment for a stock that has a price of 40 € is half a cent, and that a stock with a price above 100 € can only be transacted at prices that are multiples of 5 cents.

TABLE 2.3: Tick size table no.4 of the Federation of European Securities Exchanges.

Lower bound price (€)	Upper bound price (€)	Tick size (€)
0	9.999	0.001
10	49.995	0.005
50	99.99	0.01
100	—	0.05

In the United States, the minimal price increment has been one-eighth (1/8) of a dollar for a long time, and then became one-sixteenth (1/16) of a dollar. In particular, the format of prices was with fractions and not with decimals. Decimalization was imposed by the SEC in 2001 – see [98]. Today, except for stocks that have a price below \$1, the tick size cannot be below one cent.[21] This is Rule 612 of Reg NMS.[22]

The question of the optimal tick size for a given stock is an important question for both academics and regulators. It was addressed for instance in [58]. The revision of MiFID should address this question of tick size. We want to insist on the fact that, once tick size tables are agreed upon, the choice of the appropriate tick size category can (partially) be made by the quoted companies themselves, using stock splits or reverse splits.

---

[21] In practice, there are several mechanisms allowing for sub-penny transactions (see for instance the Retail Liquidity Program of the NYSE).

[22] In May 2015, the SEC (Securities and Exchange Commission) approved a proposal of the FINRA (Financial Industry Regulatory Authority) to widen the tick size of the stocks of small companies, over a period of two years (starting in May 2016). The goal is to test whether this widening could be beneficial to market quality.

MiFID and the tick size reduction in Europe

If the tick size of a stock is very small, i.e., if the bid-ask spread is most often equal to several ticks, then agents have no reason to post orders. There is indeed little value in entering the queue (and thus revealing information) at best prices, because one can always get higher priority by improving slightly the price. The consequence is that liquidity providers are not very active when the tick size is too small. Conversely, when the tick size is large, the bid-ask spread may be equal to 1 tick most of the time, and the volume at the best bid and ask prices are usually large. However, the cost associated with crossing the bid-ask spread may appear high.

When new actors entered the market after MiFID, they had to attract liquidity providers in order to compete with the incumbent exchanges. One way was obviously through the price of their services (mainly trading fees). Another lever was the tick size. Since MiFID imposed no rule on tick size as opposed to Reg NMS, new entrants decided to lower the tick size of some large-tick stocks. MTFs, especially Chi-X at the start, proposed lower tick sizes on their platform than on the primary venues, for a swath of stocks. Regulated markets soon followed, in order to neutralize the increase in market share the entrants had gained. This tick size reduction was certainly a good thing at the beginning, because tick sizes were indeed too large in several countries, but it soon became a tick size war (see [128]). In June 2009, the FESE launched an initiative to harmonize the different tick size regimes in Europe around a few tables. Shortly after, the MTFs agreed to stop decreasing the value of the tick size.

> From the perspective of each trading venue, strong incentives exist to undercut others in terms of tick sizes, which is not in the interest of market efficiency or the users and end investors. This might, in turn, lead to excessively reduced tick sizes in the market. Excessively granular tick sizes in securities can have a detrimental effect to market depth (i.e., to liquidity). An excessive granularity of tick sizes could lead to significantly increased costs for the many users of each exchange throughout the value chain; and have spillover costs for the derivatives exchanges clients.
>
> — The Federation of European Securities Exchanges

Trading Fees

MiFID opened the market, but competition was made real by companies which set up new trading venues. These new trading venues have competed with the incumbent exchanges mainly through the price of their service (see also the above discussion on the tick size). In particular, MTFs proposed new fee structures to attract liquidity on their platforms.[23]

Although they are all different, the fee structures of most venues are based on the principle that only executed orders are considered in the pricing. In other words, the insertion of limit orders does not lead to a charge, and the cancellation of limit orders is free.[24] Upon execution, orders are charged, the price depending on the type of order, the type of stock traded, and the volume traded by the client (on a monthly or annual basis).

As a first example, we consider Euronext. Euronext charges its best clients 0.10 basis points (bps)[25] per executed order on the CAC 40 stocks, but this price increases to 0.30 bps for the same trade by a client with a monthly activity of less than 200 million euros. Euronext also applies some additional charges of 2 cents for executed IOC and FOK orders. The second – and really different – example we consider is Turquoise, which charges 0.30 bps upon execution, if the order is a liquidity-taking one (aggressive), but offers a rebate of 0.15 to 0.26 bps[26] upon execution if the order has been providing liquidity (passive). In other words, the impatient buyer/seller removing liquidity from the book pays 0.30 bps to the platform, and the platform redistributes part of this charge to the seller/buyer whose limit order has been resting in the book. The BATS Chi-X CXE venue has a similar make/take fee structure, where liquidity providers are subsidized and liquidity takers pay for each executed order. The BATS Chi-X BXE venue has a slightly different fee structure: it charges liquidity takers only, and does not offer a rebate to liquidity providers.

It is clear that new entrants initially applied fee schemes different from those of incumbent exchanges (and not only low fees) in order to attract part of the trading flows. The optimal make/take fee structure was studied theoretically by a group of economists in [72]. They found that it could be optimal to subsidize either liquidity makers or liquidity takers, depending on the market characteristics. In Europe, venues charge both types of agents or subsidize the liquidity makers. To see a different situation, one has to cross the Atlantic Ocean.

In the United States, there is currently a dozen of lit venues, that are operated by four different companies: NASDAQ, NYSE, BATS (following the merger with Direct Edge), and CHX (Chicago Stock Exchange). Most of them,

---

[23]We consider here the trading fees only, and neither the post-trade fees (clearing and settlement), nor any fixed fees (subscription, etc.).

[24]However, many platforms apply a surcharge to orders when the order/trade ratio is too large. The threshold on Euronext for instance is 100 orders per trade, above which a surcharge of 10 cents (per order) applies.

[25]A cap and a floor in euro amount also apply, along with a monthly cap in bps.

[26]The rebate depends on the overall volume traded passively.

called regular, subsidize liquidity providers, like in Europe. However, some of them have an inverted pricing policy. They offer a rebate to the agents who remove liquidity, while there is a fee for adding liquidity (fees being paid only if the order is executed). Examples of such "inverted" venues include NASDAQ OMX BX, BATS BYX, and BATS EDGA. It is noteworthy that, in the United States, unlike in Europe, fees and rebates are usually calculated on a per share basis, not proportionally to the dollar value of the transaction.

2.2.2.3   Dark pools and hidden orders

MiFID has introduced a swath of new rules. One of the most important ones is pre-trade transparency. By imposing pre-trade transparency,[27] i.e., by requiring venues to disclose the state of order books to market participants, the regulator wants people to be aware of the current state of supply and demand before making any decisions. This may look like a natural requirement to obtain an efficient price formation process, but it has some important consequences, not to say drawbacks. In particular, large traders (traders with large orders) have to slice their orders, because posting their original large orders would reveal too much information. In practice, several waivers have been introduced. If orders are large enough (the threshold is defined in the Annex II of the MiFID regulation), they can rest in limit order books without being disclosed. This is called the large in scale (LIS) waiver. Small orders can also be hidden, but only if they are pegged to a price imported from the primary venue (often the mid-price). This is called the reference price waiver. Other waivers do exist, such as the order management facility waiver, authorizing iceberg orders (see below).

In the post-MiFID environment, most venues that are organized around a limit order book accept large hidden orders (through the LIS waiver). These large hidden orders can rest in the limit order book at any admissible price, but all visible orders have higher priority than hidden orders at the same price. Iceberg orders constitute another kind of dark liquidity executed in lit pools. In the case of an iceberg order, two sizes are chosen at insertion: a total size and a (smaller) visible size V. A limit order of size V is posted in the book and the rest lies hidden at the same price, hence the name iceberg. Once the visible part has been executed, another limit order of size V (or less if the remaining hidden portion is less than V) is posted, and the hidden portion diminishes by the same figure. Then, the mechanism repeats until execution of the total size of the order, or until the order is cancelled. Obviously, the hidden part of the iceberg order always has lower priority than any visible order at the same price. Iceberg orders can be seen as an automatic way to split and post orders. However, an iceberg order is more than that, and it is sometimes like real dark liquidity. In effect, a liquidity-taking order arriving

---

[27]Post-trade transparency is another requirement – see for instance [128] for a thorough analysis of post-trade transparency.

at the price of the iceberg order can be matched with the hidden part of the iceberg order if the size of this aggressive order is larger than the total size of visible orders at that price, and if there are no other hidden orders with higher (time) priority at the same price. Iceberg orders are usually priced differently from classical limit orders by venues. Some other orders, such as stop or stop-limit orders, are sometimes regarded as hidden orders, but this is questionable.

In addition to the lit venues organized around a limit order book with both a visible and a hidden part, some other venues, called dark pools, can be organized around a limit order book (under the LIS waiver) where all orders are hidden,[28] or can operate thanks to the reference price waiver. In the latter case, orders can be smaller than the LIS threshold, but they are executed at the bid, mid, or ask price of the regulated market. Such dark pools are set up by the companies operating lit pools, or by other financial actors: brokers with their crossing networks (BCNs), banks setting their own dark pool – e.g., UBS MTF in Table 2.1. Market participants use them in order to avoid disclosing information (for reducing market impact), but also to avoid crossing the bid-ask spread in the particular case of mid-point dark pools.

The rationale for the existence of mid-point dark pools, or other dark pools operating thanks to the reference price waiver, is that the transactions occur at a fair price, because the price is imported from the primary venue, where the price formation process is assumed to be efficient. However, because of fragmentation and dark liquidity, the price formation process may not be as efficient as initially assumed. Dark pools raise in fact a lot of questions, especially because the price formation process has been deeply modified by the presence of high-frequency traders.

2.2.2.4 High-frequency trading

High-frequency trading (HFT) is often described as being evil. In fact, high-frequency trading is very diverse. Some high-frequency trading strategies can be assimilated as price manipulation, and they are indeed detrimental to the price formation process. However, some high-frequency traders, often referred to as high-frequency market makers, play a major role in the current trading environment. They act indeed as liquidity-providing arbitrageurs across the different platforms quoting the same stock. Therefore, they make it possible for the information to circulate almost instantaneously from one pool to the others. In other words, high-frequency market makers are essential to solve what L. Harris called in [98] the problem of the "two mostly incompatible competitions": the competition between traders who buy and sell shares, and

---

[28] Off-market trades have a long history. Historically, large orders were often executed on the upstairs markets in order to reduce market impact. Modern dark pools based on the LIS waiver could be viewed as the electronic equivalent of the old upstairs markets.

the competition between platforms for organizing the competition between traders. High-frequency trading is in fact the price to pay to have at the same time low trading fees (because of competition between venues) and a situation as close as possible, from an informational point of view, to a situation where all market participants meet at the same place.

Although high-frequency trading plays an important role in the current competitive structure, it also raises many questions. Some market participants are indeed unwilling to trade with high-frequency traders because of the potential adverse selection, not to say because they fear gaming. Bad practices such as layering or spoofing have been described. Quote stuffing is another one, and platforms try to reduce these practices through additional costs after a given threshold of the order/trade ratio is reached. To avoid information leakage towards high-frequency traders, utilitarian traders tend to trade more and more OTC or on platforms without pre-trade transparency.[29] This is a problem for the quality of the price formation process. We hope that MiFID 2 will improve this environment.

2.3 Bond markets

2.3.1 Introduction

Historically, stock exchanges were marketplaces to trade securities, in which bonds and stocks were bought and sold in a similar way. In the 19th century for instance, the Paris Bourse and the London Stock Exchange were the main places to trade stocks and bonds issued for financing projects from all over the world. Bond trading and stock trading have shared a common history until the progressive worldwide adoption, in the 1980s and 1990s, of order-driven trading systems on the equity side.[30] As opposed to stocks, most bonds continue today to be traded OTC in a dealer (or quote-driven) market, that is, without order books.

There are many differences between order-driven markets and quote-driven markets. By definition, in a quote-driven market, there is a formal distinction between two types of market participants:[31] dealers and clients. Dealers (in general, investment banks) are market makers and liquidity providers: they

---

[29]However, some high-frequency traders may try to manipulate prices on the primary venue to make money out of the liquidity resting in mid-point dark pools – see [128].

[30]In the United Kingdom, the main limit order book of the LSE, called the SETS (Stock Exchange electronic Trading Service), was launched only in 1997.

[31]This is also true in stock markets when there are specialists. However, trading is organized around a limit order book (for most stocks).

propose bid and offer quotes for the securities (here bonds). Clients (institutional investors, wealth management companies, hedge funds, etc.) trade with them. In particular, clients do not trade with other clients. On the contrary, in an order-driven market, all agents can send their orders to buy and sell securities (during auctions or during the continuous phase of trading), and the price is formed from the interaction between all buyers and sellers. In the former case, there is no transparency, and there is also no anonymity, unlike what happens in the latter market structure. In particular, dealers can discriminate clients in a quote-driven market because of this lack of anonymity.[32]

Although still mainly quote-driven, bond markets have recently undergone many changes because of a partial electronification, and because of the regulation on banks that imposed restrictions on the inventories of dealers. The next paragraphs aim at explaining why bonds are not traded like stocks, and at describing how the bond market works. As for stocks, our focus is on the European environment.

2.3.2   Bond markets and liquidity

Electronic order books are now the standard for a lot of asset classes: stocks, foreign exchanges, etc. Although many historically dealer-driven markets have eventually adopted the order-driven paradigm, this is not the case for most bond markets. In fact, it is hard to imagine such a market organization for bonds, except for the government bonds of the most important countries, and for the most liquid corporate bonds.

Let us compare stocks and corporate bonds for instance. There is usually one stock for a given company while there are often dozens of bonds for the same company, corresponding to different maturities, coupons, and seniorities.[33] A natural consequence is that most corporate bonds are illiquid. The Securities Industry and Financial Markets Association (SIFMA) estimates indeed that the total market value of stocks is twice as large as the market value of corporate bonds, but that there are more than six times more listed corporate bonds than listed stocks. Regarding liquidity, MarketAxess Research (see [144]) estimates that, in 2012, 38% of the 37,000 TRACE-eligible bonds did not trade, not even once, and that only 1% of these 37,000 bonds traded every day. Government bonds represent a more liquid side of the bond world, but there is still, for most countries, a huge list of bonds with different characteristics.

---

[32] Anonymity is also a problem on stock markets because the toxic liquidity of some high-frequency traders cannot be easily identified.

[33] Somehow, bonds should not be compared to stocks, but instead to options. In both cases, there are indeed, for each underlying, a large number of products with different maturities and different characteristics.

Bonds are not standardized products. They have various features and various maturities corresponding to the various needs of companies and governments. Bonds are not easily traded by individual investors (stocks usually have low prices whereas bonds have a large notional). Furthermore, many (institutional) investors have and will continue to have buy-and-hold strategies with respect to bonds. All these reasons explain why bond markets are largely illiquid. Moreover, even if bonds were made more divisible and standardized products, the fact that many investors buy bonds for the coupons, not for potential capital gains, and therefore never go to the secondary market, makes it difficult to imagine the emergence of an order-driven market with order books similar to those of equity markets. In fact, most practitioners (see [144]) believe that the lack of liquidity on the secondary market will remain in the future, and that the attempts to create limit order books cannot be successful.

Quote-driven markets can work well for illiquid assets. Nevertheless, for this market structure to be efficient, dealers have to maintain large inventories (long and short). The job of dealers is to make the market, that is, answering requests from clients and proposing bid and offer prices for different volumes, including large volumes – and making money out of the bid-ask spread they quote. As clients often trade in the same direction at the same time, for bond markets to work well, dealers must be able to carry large inventories. However, it is already clear that the Basel III capital requirements deter investment banks from holding such large inventories. In particular, it appears that there is recurrently a sharp decrease in the liquidity of corporate bonds, because banks drastically reduce their inventories in order to be in a better configuration in front of the regulator.

Bond markets today, especially the European corporate bond market, are in turmoil. The classical quote-driven structure is less and less compatible with the constraints affecting the investment banks (the dealers). However, an efficient market structure organized around a central limit order book would only be suited for a few liquid government and corporate bonds. The future of bond markets is in fact not very clear, except on one point: even if trading remains largely OTC, voice trading will progressively – although not entirely – be replaced by electronic platforms.

2.3.3   Electronification of bond trading

Bonds are traditionally divided into two types: government bonds and non-government bonds.[34] For both types of bonds, we distinguish two different market segments: the dealer-to-client (D2C) segment, where transactions

---

[34]We use the term corporate bonds for non-government bonds in general, including therefore the bonds issued by financial institutions.

occur between dealers and clients, and the inter-dealer broker (IDB) segment. The characteristics of the D2C and IDB segments are specific for each of the two types of bonds.

2.3.3.1 Corporate bonds

Technological innovation occurs in the D2C market segment that is increasingly electronifying, whereas trades in the IDB market segment for corporate bonds remain almost entirely executed via voice. Several types of platforms have emerged. The most important and successful ones are multi-dealer-to-client (MD2C) trading platforms.

Examples of such MD2C platforms are those proposed by Bloomberg Fixed Income Trading (FIT), Tradeweb and MarketAxess – Bloomberg's platform being the most widely used. These platforms clearly dominate the landscape of electronic trading for corporate bonds. With these platforms, there is still a distinction between dealers and clients, but clients can simultaneously send a request for quotes (RFQ) to several dealers who have streamed prices to the platforms. Another kind of MD2C electronic platform that is used by many buy-siders (especially small ones) consists of executable quotes, but only for odd lots. Single-dealer electronic platforms have also emerged to replace the telephone. Crossing systems, and platforms proposing all-to-all limit order books also exist. They try to revolutionize the classical distinction between dealers who provide liquidity and clients who take liquidity, but they are still very rarely used in practice.

An interesting study carried out by McKinsey & Company (see [144]) shows that the current market structure is far from being stabilized, given the different viewpoints of market participants on the future organization of the corporate bond market. It is clear that the shift from voice trading to electronic platforms will go on. It is believed by a majority of practitioners that the market will remain for years a dealer-driven one, with MD2C RFQ platforms holding the lion's share of electronic trading. Moreover, the evolution of financial regulation will have a lot of influence on the market structure and the role of the different participants. As mentioned above, Basel III partially prevents dealers from playing their traditional role. This might encourage all-to-all platforms, and give a new role to large asset managers, as specialists on these platforms. Furthermore, the question of pre-trade transparency for corporate bonds is raised for the definition of future regulations, and the European landscape may change to apply new rules.

The typical process to buy/sell a bond on an MD2C platform

By using MD2C platforms, buy-siders can simultaneously send requests for quotes (RFQ) to several dealers on a specific bond. These requests can be sent with the willingness to buy or sell a bond, or simply to catch information. The typical process is the following:

1. The client connects to the MD2C platform and sees the bid and offer prices for a given bond. These prices, streamed by dealers, correspond to prices for a given reference size. They are not firm/executable prices.

2. The client selects dealers (up to 6 dealers on Bloomberg FIT for instance), and sends one RFQ through the platform to these dealers, with a precise volume, and the side (buy or sell order).

3. Requested dealers can answer a price to the client for the transaction. Dealers know the identity of the client (no anonymity), and the number of dealers requested (the degree of competition). However, they do not see the prices streamed by other dealers. They only see a composite price at the bid and offer, based on some of the best streamed prices. This composite is called the CBBT price in the case of Bloomberg FIT .

4. The client receives progressively the answers to the RFQ. He can deal at any time with the dealer who has proposed the best price (or one of them if two dealers proposed the same price), or decide not to trade.

5. Each dealer knows whether a deal was done (with him, but also with another dealer – without knowing the identity of this dealer), or not. When a transaction occurs, the best dealer usually knows the price(s) proposed by his competitor(s) (if there is any).

2.3.3.2 Government bonds

The shift from voice trading to electronic trading is more advanced for government bonds than for corporate bonds. In particular, as opposed to the IDB market for corporate bonds that has not evolved in its structure, a substantial part of the IDB market for government bonds is electronified. The leading actor on this segment is MTS, which proposes both an inter-dealer electronic platform and limit order books, the latter being still very marginally used.

On the D2C market, voice trading is progressively replaced by MD2C platforms, single-dealer platforms, and all-to-all limit order books. As for corporate bonds, MD2C platforms dominate. The main actors on this segment are Tradeweb, Bloomberg, BondVision, and MarketAxess.

2.4   Conclusion

The way stocks are traded is constantly evolving because of both new technologies and regulatory changes. The fragmentation of liquidity is a major feature of stock markets today. It has led to the development of new trading strategies, and new tools, such as smart order routers. In terms of financial modeling, the new stock trading environment has pushed academic research towards new topics, such as optimal execution, optimal placement of orders, or the determination of optimal market making strategies. MiFID 2 will certainly change the trading environment again, but it is hard to guess what will be, and not be, included in the revision of the regulation. We bet that the problems tackled in this book will be even more relevant after the implementation of MiFID 2.

As far as bonds are concerned, future regulations may deeply change the market structure by imposing pre-trade transparency. We bet that research on block trade pricing (see Chapter 8) and market making (see Chapter 11) will be key for the future, especially for dealers to automate quote streaming and bond trading.

Part II

Optimal Liquidation

Chapter 3

The Almgren-Chriss framework

Suum cuique decus posteritas rependit.

— Tacitus

What is the optimal pace to build or unwind a position? In this chapter, we present the modeling framework introduced by Almgren and Chriss to schedule execution, and therefore to answer this important question. We first present the ideas that made the Almgren-Chriss approach a success. Then, we build a general model – inspired by the initial models presented in [8, 9] – for determining the optimal trading curve associated with buying or selling a large block of shares. This model is presented both in continuous time, for introducing the general framework, and in discrete time, because the discrete-time formulation is useful when it comes to numerics.

3.1   Introduction

The problems faced by cash traders and brokers have long been ignored by both researchers in financial mathematics and economists. The first academic paper dealing with the optimal execution of a block of shares was only published in 1998, by Bertsimas and Lo (see [21]). In this first paper, Bertsimas and Lo modeled the execution price of a single trade as a function of the current market state and the number of shares actually traded. Given such an execution cost function, they proposed a way to "optimally" split a large order into small orders to be executed over a given time window. In their approach, optimality means that the expected cost of execution is minimized.

Bertsimas and Lo definitely blazed a trail, but they did not take into account any risk concerns. In particular, they did not take into account that the market price may move over the course of the execution process.[1] Therefore, their optimal strategies often boil down to breaking up the large order evenly into small orders to be executed at constant speed. In 1999, a year after Bertsimas and Lo, Almgren and Chriss proposed in [8] a model taking into account both the expected cost of execution and the risk that the price

---

[1] Taking price risk into account is very important for financial intermediaries, such as brokers, but far less important for fund managers who execute in-house their own orders.

would move over the course of the execution process. Large orders are split into smaller ones, that are executed progressively over a given time window. However, the optimal scheduling of the execution process is not as simple as in [21]. A trader (or an algorithm) executing fast pays high execution costs because liquidity is limited, but a slow execution exposes to possible adverse price fluctuations. Therefore, there is a trade-off between execution costs on the one hand, and price risk on the other hand. This trade-off, which is central in the literature on optimal execution, leads to nontrivial optimal execution strategies.

Almgren and Chriss are unanimously considered, and rightly so, as the pioneers of the early literature on optimal execution. The above trade-off could be found almost at the same time in other sources (see for instance the book of Grinold and Kahn [78]), but (at least) three features of the model proposed by Almgren and Chriss in [8], and then in [9], led to its success. First, they derived a closed-form formula for the optimal execution strategy. Second, the optimal execution strategy they derived takes the form of a trading curve that can be computed before the start of the execution process. Third, they proposed a simple way to model market impact and execution costs.

Almgren and Chriss indeed considered a simple model for the market impact. They separated it into two parts: in their modeling framework, market impact is either instantaneous or permanent. The initial idea behind this decomposition is a very simple, and in fact very simplistic, one. When a market order is sent, the price of the transaction depends on the size of the order. This price takes account of execution costs, such as those associated with the bid-ask spread, or linked to the limited available liquidity at each price in the order book. It is the instantaneous part of the market impact in the terminology used by Almgren and Chriss: everything works as if the market price was impacted by the transaction and returned instantaneously to its initial state. In addition to this first component, Almgren and Chriss assumed that the market price of a stock is permanently impacted by trades. Sending a market order to buy (sell) shares pushes the price up (down), and there is no resilience in the model.

Modeling market impact is important to better understand the price formation process and the impact of the market microstructure on trading. The price impact of trades has been studied by economists, mathematicians, and even physicists – or more precisely econophysicists. Their goal is to understand the underlying mechanisms of the impact of trades on prices, and to quantify the magnitude of this impact and its dynamics. The market impact model proposed by Almgren and Chriss is far from being realistic. The market price is obviously (partially) resilient, and the dynamics of the impact is not as simple as one part that vanishes instantaneously, and another part that stays forever.

Nevertheless, the model introduced by Almgren and Chriss is a great model for reasons related to the way execution algorithms are built. Most execution algorithms used by banks and brokerage companies are built with at least two layers: a strategic one and a tactical one. The strategic layer is in charge of scheduling the execution process; in other words, it only computes the optimal trading curve. The tactical layer is more complex: it deals with the routing of orders across platforms/venues and the micromanagement of orders. Although the model initially built by Almgren and Chriss was a model for the entire execution process (using market orders only), it should be seen as a scheduling model, that is, a model for the strategic layer of execution algorithms. With this modern viewpoint, the instantaneous component of market impact does not quantify the impact of market orders, but instead the average cost of trading that cannot be explained by the unavoidable component of market impact (which is the modern understanding of the permanent market impact in the Almgren-Chriss model). In particular, it depends on the ability of the tactical layer to obtain good prices across venues, over short periods of time, using all kinds of orders. For this reason, in this book we prefer calling "execution costs" this instantaneous form of market impact. Furthermore, we can include in these execution costs some transaction costs, such as fees paid to the trading platforms, stamp duties, or financial transaction taxes. Therefore, a good scheduling model is a bottom-up model, with execution cost and market impact functions estimated in a top-down manner.

The model we present in this chapter is a generalized version of the initial model of Almgren and Chriss. Instead of focusing on the case of quadratic execution costs, we consider a more general case. We first present a continuous-time version of this model. Then, we present a discrete-time version of this model, which plays an important part as far as numerical applications are concerned.

3.2 A generalized Almgren-Chriss model in continuous time

3.2.1 Notations

We consider a trader with a single-stock portfolio. At time $t = 0$, his position is denoted by q_0. If $q_0 > 0$, then the portfolio contains q_0 shares while, if $q_0 < 0$, then the trader has a short position in the stock, of $-q_0$ shares.

We consider the problem of unwinding this portfolio over a time window $[0, T]$. The trader's position over the time interval $[0, T]$ is modeled by the process $(q_t)_{t \in [0,T]}$.

The dynamics of this process is

$$dq_t = v_t dt, \tag{3.1}$$

where $(v_t)_{t\in[0,T]}$ is a progressively measurable control process satisfying the unwinding constraint $\int_0^T v_t dt = -q_0$, and the additional technical condition $\int_0^T |v_t| dt \in L^\infty(\Omega)$.

We denote by $\mathcal{A}$ the set of such admissible controls:

$$\mathcal{A} = \left\{ (v_t)_{t\in[0,T]} \in \mathbb{H}^0(\mathbb{R}, (\mathcal{F}_t)_t), \int_0^T v_t dt = -q_0, \int_0^T |v_t| dt \in L^\infty(\Omega) \right\}.$$

At time t, v_t stands for the trading velocity, that is, the (instantaneous) trading volume.

The mid-price[2] of the stock is modeled by the process $(S_t)_{t\in[0,T]}$. Trades impact this price: this is the permanent component of market impact described in the introduction. We assume a linear dependence of the form

$$dS_t = \sigma dW_t + k v_t dt, \tag{3.2}$$

where σ is a positive constant called the arithmetic volatility of the stock, and k is a nonnegative parameter modeling the magnitude of the permanent market impact.

Normal dynamics vs. log-normal dynamics

In most models in the literature on optimal execution, prices are assumed to be normally distributed. This contrasts with the basic framework of option pricing, where price returns are normally distributed, and prices follow therefore a log-normal dynamics.

The reason for such a choice is that it simplifies the model. The main problem with normally distributed prices is that prices can become negative. However, execution problems are usually raised over a few minutes or hours. Therefore, the probability that the price reaches negative values is very low. Moreover, for standard values of the volatility parameter, the dynamics of log-normal prices is very similar to the dynamics of normally distributed prices over short periods of time.

---

[2]There is in fact no need to consider the mid-price in particular. The bid price, the ask price, or even the price of the last trade would be fine, as long as the bid-ask bounce is filtered out of the estimation of the volatility.

Because of execution costs, the price obtained by the trader at time t is not S_t, but instead a price that depends on the volume he trades, and on the market volume at that time.

To model these execution costs, we first introduce the market volume process $(V_t)_{t \in [0,T]}$, which represents the volume traded by other agents. We assume that it is a deterministic,[3] continuous, positive, and bounded process.

The price obtained by the trader for each share at time t is of the form $S_t + g\left(\frac{v_t}{V_t}\right)$, where g is an increasing function satisfying $g(0) = 0$. If the trader buys (sells) shares, then the trading price is higher (lower) than the mid-price.

Instead of working with the function g, we introduce the execution cost function L that is simply $L(\rho) = \rho g(\rho)$.

The trader buys shares with cash, and gets cash when he sells shares. We denote by $(X_t)_t$ the cash account process modeling the amount of cash on the trader's account. Given the above assumptions, its dynamics is given by

$$
\begin{aligned}
dX_t &= -v_t \left(S_t + g\left(\frac{v_t}{V_t}\right) \right) dt \\
&= -v_t S_t dt - V_t L\left(\frac{v_t}{V_t}\right) dt.
\end{aligned}
\tag{3.3}
$$

The assumptions on the function $L : \mathbb{R} \to \mathbb{R}$ are the following:

- (H1) No fixed cost, i.e., $L(0) = 0$,
- (H2) L is strictly convex, increasing on $\mathbb{R}_+$ and decreasing on $\mathbb{R}_-$,[4]
- (H3) L is asymptotically super-linear, i.e., $\lim_{|\rho| \to +\infty} \frac{L(\rho)}{|\rho|} = +\infty$.

Assumption (H1) means that no execution costs are incurred when there is no transaction, hence no fixed cost. Assumption (H2) means that there is always a cost to trade (both when selling and when buying), and that this cost is more than just a linear cost related to the bid-ask spread or a stamp duty. Assumption (H3) is a technical assumption.

In practical examples, we often choose for L a strictly convex power function (i.e., $L(\rho) = \eta |\rho|^{1+\phi}$ with $\phi > 0$), or a function of the form $L(\rho) = \eta |\rho|^{1+\phi} + \psi |\rho|$ with $\phi, \psi > 0$, where the additional term $\psi |\rho|$ models proportional costs such as the bid-ask spread, the fees paid to the venue,

---

[3] Although this assumption may seem odd because volume cannot be exactly predicted, it corresponds to the way many algorithms work in practice, with static volume curves computed using historical data (see Chapter 4).

[4] (H2) can be replaced by the assumption of strict convexity and nonnegativity, because of (H1).

and/or a stamp duty.[5] The initial Almgren-Chriss models correspond to a quadratic function $L(\rho) = \eta\rho^2$.

Why should permanent market impact be linear?

In our model, we assume that the permanent component of market impact is linear. This assumption guarantees the absence of dynamic arbitrage, as defined by Gatheral in [74] and recalled below (see also Huberman and Stanzl [109]).

Imagine indeed that the permanent market impact is modeled by a function $\kappa(\cdot)$. Let us also suppose that there is no execution cost. Then the dynamics of (q_t, S_t, X_t) is

$$\begin{cases} dq_t &=& v_t dt, \\ dS_t &=& \sigma dW_t + \kappa(v_t)dt, \\ dX_t &=& -v_t S_t dt. \end{cases}$$

There is a dynamic arbitrage if there exist $t_1 < t_2$, and a progressively measurable process $(v_t)_t$ such that the following three conditions are satisfied:

- $\int_{t_1}^{t_2} |v_t| + |\kappa(v_t)| dt \in L^\infty(\Omega)$,

- $\int_{t_1}^{t_2} v_t dt = 0$,

- $\mathbb{E}[X_{t_2}|\mathcal{F}_{t_1}] > X_{t_1}$.

In other words, a dynamic arbitrage corresponds to a round trip strategy on the stock that is profitable on average.

In fact, $\kappa(\cdot)$ linear is the only possible choice to guarantee the absence of dynamic arbitrage. To see it, let us consider an interval $[t_1, t_2]$, and two scalars α and β of the same sign. The round trip strategy

$$v_t = \begin{cases} \alpha & \text{if } t \in [t_1, \tau(\alpha, \beta)], \\ -\beta & \text{if } t \in [\tau(\alpha, \beta), t_2], \end{cases}$$

with $\tau(\alpha, \beta) = \frac{\alpha t_1 + \beta t_2}{\alpha + \beta}$, gives

$$X_{t_2} = X_{t_1} - \int_{t_1}^{t_2} v_t S_t dt = X_{t_1} + \int_{t_1}^{t_2} (q_t - q_{t_1})\sigma dW_t + \int_{t_1}^{t_2} (q_t - q_{t_1})\kappa(v_t)dt.$$

---

[5]In this book, the term $\psi|\rho|$ is called the bid-ask spread component.

Therefore, using the fact that $q_{t_1} = q_{t_2}$, we have

$$
\begin{aligned}
\mathbb{E}[X_{t_2}|\mathcal{F}_{t_1}] &= X_{t_1} + \int_{t_1}^{t_2} (q_t - q_{t_1})\kappa(v_t)dt \\
&= X_{t_1} + \int_{t_1}^{\tau(a,b)} (q_t - q_{t_1})\kappa(v_t)dt + \int_{\tau(a,b)}^{t_2} (q_t - q_{t_2})\kappa(v_t)dt \\
&= X_{t_1} + \int_{t_1}^{\tau(a,b)} \alpha(t - t_1)\kappa(\alpha)dt + \int_{\tau(a,b)}^{t_2} \beta(t_2 - t)\kappa(-\beta)dt \\
&= X_{t_1} + \frac{1}{2}\alpha \left(\frac{\beta}{\alpha + \beta}\right)^2 (t_2 - t_1)^2\kappa(\alpha) \\
&\quad + \frac{1}{2}\beta \left(\frac{\alpha}{\alpha + \beta}\right)^2 (t_2 - t_1)^2\kappa(-\beta) \\
&= X_{t_1} + \frac{1}{2}\frac{\alpha\beta}{(\alpha + \beta)^2}(t_2 - t_1)^2 (\beta\kappa(\alpha) + \alpha\kappa(-\beta)).
\end{aligned}
$$

If there is no dynamic arbitrage, then we must have $\mathbb{E}[X_{t_2}|\mathcal{F}_{t_1}] \leq X_{t_1}$. This should be true for any choice of α and β corresponding to a round trip. Therefore,

$$\forall \alpha, \beta \in \mathbb{R}, \alpha\beta > 0 \Rightarrow \beta\kappa(\alpha) + \alpha\kappa(-\beta) \leq 0.$$

Replacing α by $-\beta$, and β by $-\alpha$, we see that we must have

$$\forall \alpha, \beta \in \mathbb{R}, \alpha\beta > 0 \Rightarrow \alpha\kappa(-\beta) + \beta\kappa(\alpha) \geq 0.$$

Therefore,

$$\forall \alpha, \beta \in \mathbb{R}, \alpha\beta > 0 \Rightarrow \beta\kappa(\alpha) = -\alpha\kappa(-\beta).$$

In particular, by setting $\alpha = \beta$, we see that $\kappa(\cdot)$ is an odd function on $\mathbb{R}^*$. Now, if $\alpha \neq 0$ and $\beta = \text{sign}(\alpha)$, we have

$$\forall \alpha \in \mathbb{R}^*, \kappa(\alpha) = -\alpha\text{sign}(\alpha)\kappa(-\text{sign}(\alpha)) = \alpha\kappa(1).$$

We now need to prove that $\kappa(0) = 0$. For this purpose, we assume the contrary and we consider the following round trip strategy:

$$
v_t = \begin{cases}
\kappa(0) & \text{if } t \in [t_1, t_1 + \frac{\tau}{3}], \\
0 & \text{if } t \in [t_1 + \frac{\tau}{3}, t_1 + \frac{2\tau}{3}], \\
-\kappa(0) & \text{if } t \in [t_1 + \frac{2\tau}{3}, t_2],
\end{cases}
$$

where $\tau = t_2 - t_1$.

In that case,

$$
\begin{aligned}
\mathbb{E}[X_{t_2}|\mathcal{F}_{t_1}] &= X_{t_1} + \int_{t_1}^{t_2} (q_t - q_{t_1})\kappa(v_t)dt \\
&= X_{t_1} + \int_{t_1}^{t_1+\frac{\tau}{3}} \kappa(0)(t - t_1)\kappa(\kappa(0))dt \\
&+ \int_{t_1+\frac{\tau}{3}}^{t_1+\frac{2\tau}{3}} \frac{\tau}{3}\kappa(0)^2 dt + \int_{t_1+\frac{2\tau}{3}}^{t_2} \kappa(0)(t_2 - t)\kappa(-\kappa(0))dt \\
&= X_{t_1} + \kappa(0)^2\frac{\tau^2}{9} > X_{t_1}.
\end{aligned}
$$

The absence of dynamic arbitrage implies that $\kappa(0) \neq 0$ is impossible.

The conclusion is
$$\forall \alpha \in \mathbb{R}, \kappa(\alpha) = \alpha\kappa(1).$$

Conversely, if $\kappa(v) = kv$, then for any process $(v_t)_t$ satisfying the following two conditions:

- $\int_{t_1}^{t_2} |v_t|dt \in L^\infty(\Omega)$,
- $\int_{t_1}^{t_2} v_t dt = 0$,

we have

$$
\mathbb{E}[X_{t_2}|\mathcal{F}_{t_1}] = X_{t_1} + \int_{t_1}^{t_2} k(q_t - q_{t_1})v_t dt = X_{t_1} + \frac{k}{2}(q_{t_2} - q_{t_1})^2 = X_{t_1}.
$$

Therefore, there is no dynamic arbitrage with a linear function $\kappa(\cdot)$.

Obviously, we choose $\kappa(v) = kv$ with $k \geq 0$ to model the fact that buying (selling) pushes the price up (down).

Although this is only rarely done in the literature, it is possible to consider that the permanent component of market impact depends on the number of shares already traded. For instance, a model where the price dynamics is

$$dS_t = \sigma dW_t + f(|q_0 - q_t|)v_t dt,$$

with f a positive function (usually decreasing), does not lead to any dynamic arbitrage (see [80]).

3.2.2 The optimization problem

Our goal is to find an optimal strategy $(v_t)_t \in \mathcal{A}$ to liquidate the portfolio. For that purpose, we need to decide what optimality means. The framework proposed by Bertsimas and Lo in [21] corresponds to maximizing $\mathbb{E}[X_T]$, or more exactly finding a strategy which maximizes $\mathbb{E}[X_T]$. In [8, 9], Almgren and Chriss considered instead a mean-variance criterion. They proposed to maximize an expression of the form $\mathbb{E}[X_T] - \frac{\gamma}{2}\mathbb{V}[X_T]$, where γ is a positive constant. Here, to remain into a classical economic framework, we consider an expected utility criterion. The utility function we consider is a CARA (Constant Absolute Risk Aversion) utility function, that is, an exponential utility function. In other words, our objective function is of the form

$$\mathbb{E}\left[-\exp(-\gamma X_T)\right],$$

where γ is a positive constant, called the absolute risk aversion coefficient of the trader.

To solve this problem, we first restrict the class of admissible controls $(v_t)_t$ to deterministic ones. Then, we show that no stochastic admissible control can do better than the best deterministic one. Almgren and Chriss, in their seminal papers, only considered deterministic strategies. The first proof that there is no need to consider stochastic strategies in the case of a CARA utility function is due to Schied et al. (see [162]). We prove this result in Section 3.2.4.

3.2.3 The case of deterministic strategies

3.2.3.1 A unique optimal strategy

In this section, we restrict liquidation strategies to deterministic ones. The main consequence of this assumption is that a strategy can be represented by a function $t \mapsto q(t)$,[6] computed at the beginning of the execution process. In particular, the strategy does not depend on the evolution of the price.

Formally, we consider the restricted set of admissible control processes $\mathcal{A}_{\text{det}}$ defined by

$$\mathcal{A}_{\text{det}} = \left\{(v_t)_{t \in [0,T]} \in \mathcal{A}, \forall t \in [0,T], v_t \text{ is } \mathcal{F}_0 - \text{measurable}\right\}.$$

Given a strategy $(v_t)_{t \in [0,T]} \in \mathcal{A}$, let us compute the final value of the cash process X_T.

---

[6]The graph of such a function (and sometimes the function itself) is called a trading curve.

By using Eqs. (3.1), (3.2), and (3.3) and an integration by parts, we get

$$
\begin{aligned}
X_T &= X_0 - \int_0^T v_t S_t dt - \int_0^T V_t L\left(\frac{v_t}{V_t}\right) dt \\
&= X_0 + q_0 S_0 + \int_0^T k v_t q_t dt + \sigma \int_0^T q_t dW_t - \int_0^T V_t L\left(\frac{v_t}{V_t}\right) dt \\
&= X_0 + q_0 S_0 - \frac{k}{2} q_0^2 + \sigma \int_0^T q_t dW_t - \int_0^T V_t L\left(\frac{v_t}{V_t}\right) dt. \quad (3.4)
\end{aligned}
$$

If $(v_t)_{t \in [0,T]} \in \mathcal{A}_{\text{det}}$, then the final value of the cash process X_T is normally distributed with mean[7]

$$
\mathbb{E}[X_T] = X_0 + q_0 S_0 - \frac{k}{2} q_0^2 - \int_0^T V_t L\left(\frac{v_t}{V_t}\right) dt, \quad (3.5)
$$

and variance

$$
\mathbb{V}[X_T] = \sigma^2 \int_0^T q_t^2 dt. \quad (3.6)
$$

The mean of X_T can be decomposed into three parts:

$$
\mathbb{E}[X_T] = \underbrace{X_0 + q_0 S_0}_{\text{MtM value}} - \underbrace{\frac{k}{2} q_0^2}_{\text{perm. m. i.}} - \underbrace{\int_0^T V_t L\left(\frac{v_t}{V_t}\right) dt}_{\text{execution costs}}. \quad (3.7)
$$

The first term, $X_0 + q_0 S_0$, is the Mark-to-Market (MtM) value of the portfolio at time 0, i.e., the cash value of the portfolio if it were possible to liquidate the shares instantaneously at their current market price. The second term corresponds to costs coming from the permanent component of market impact. If the unwinding process consists in selling shares, then we sell shares at a price that (on average) decreases over the course of the selling process. Similarly, if the unwinding process consists in buying shares, then we buy shares at a price that (on average) increases over the course of the buying process. In both cases, it leads to a discount term $-\frac{k}{2} q_0^2$. It is interesting to notice that this term does not depend on the strategy $(v_t)_t$. In other words, the costs coming from the permanent component of market impact cannot be avoided. In contrast, $(v_t)_t$ appears in the third term of Eq. (3.7), which corresponds to the execution costs.

Because L is assumed to be convex, the last term of Eq. (3.5) is minimal when v_t is proportional to V_t (this is the strategy proposed by Bertsimas and Lo in [21]). This strategy is not optimal in our framework because we take the variance into account. The variance of X_T is an increasing function of

---

[7]X_T can be equal to $-\infty$. In that case, we still say that X_T follows a Gaussian distribution, but the mean is $-\infty$.

the volatility parameter σ, and it depends on the strategy through the term $\int_0^T q_t^2 dt$. To minimize this variance term, the trader must quickly liquidate. The trade-off between execution costs and price risk is here a trade-off between minimizing the last term of Eq. (3.5), and minimizing the variance of X_T.

By using Eq. (3.5), Eq. (3.6), and the formula for the Laplace transform of a Gaussian variable (see Theorem A.1 of Appendix A), we can compute the value of the objective function:

$$
\begin{aligned}
\mathbb{E}\left[-\exp\left(-\gamma X_T\right)\right] &= -\exp\left(-\gamma \mathbb{E}\left[X_T\right] + \frac{1}{2}\gamma^2 \mathbb{V}\left[X_T\right]\right) \\
&= -\exp\left(-\gamma\left(X_0 + q_0 S_0 - \frac{k}{2}q_0^2\right)\right) \\
&\quad \times \exp\left(\gamma\left(\int_0^T V_t L\left(\frac{v_t}{V_t}\right)dt + \frac{\gamma}{2}\sigma^2 \int_0^T q_t^2 dt\right)\right).
\end{aligned}
$$

As a consequence, the problem boils down to finding a control process $(v_t)_{t\in[0,T]} \in \mathcal{A}_{\text{det}}$ minimizing

$$
\int_0^T V_t L\left(\frac{v_t}{V_t}\right)dt + \frac{\gamma}{2}\sigma^2 \int_0^T q_t^2 dt. \tag{3.8}
$$

Because $v_t = \frac{dq_t}{dt}$ is the first derivative of q_t, and $\int_0^T |v_t| dt < +\infty$, the function $q : t \mapsto q_t$ is in fact absolutely continuous. Therefore, the problem boils down to a variational problem (a Bolza problem, to be precise). We need to find the minimizers of the functional J defined by

$$
J(q) = \int_0^T \left(V_t L\left(\frac{q'(t)}{V_t}\right) + \frac{1}{2}\gamma\sigma^2 q(t)^2\right) dt, \tag{3.9}
$$

over the set of absolutely continuous functions $q \in W^{1,1}(0,T)$ satisfying the constraints $q(0) = q_0$ and $q(T) = 0$.

The following theorem states the main results:

Theorem 3.1. *There exists a unique minimizer q^* of the function J over the set $\mathcal{C} = \{q \in W^{1,1}(0,T), q(0) = q_0, q(T) = 0\}$. Furthermore, q^* is a monotone function:*

- *if $q_0 \geq 0$, q^* is a nonincreasing function of time,*
- *if $q_0 \leq 0$, q^* is a nondecreasing function of time.*

Proof. *The existence of a minimizer q^* is guaranteed by classical results on Bolza problems (see for instance [157], Theorem 6.12 of [35], or Theorem B.3 in Appendix B).*

Uniqueness is straightforward. If indeed $q_1, q_2 \in \mathcal{C}$ are distinct functions such that $J(q_1) = J(q_2) = \min_{\mathcal{C}} J$, then because L is a convex function and $q \mapsto q^2$ a strictly convex function, $J\left(\frac{q_1 + q_2}{2}\right) < \frac{J(q_1) + J(q_2)}{2} = \min_{\mathcal{C}} J$. This contradicts the minimality of $J(q_1)$ and $J(q_2)$. Therefore, there is a unique minimizer.

For proving that q^* is a monotone function, we consider the case $q_0 \geq 0$. The other case is similar.

We introduce the strategy q^{**} defined by $q^{**}(t) = \inf_{s \leq t} \max(q^*(s), 0)$. We have the following:

- *$q^{**} \in W^{1,1}(0, T)$, with $q^{**'}(t) = q^{*'}(t) 1_{\{q^*(t) > 0, q^*(t) = q^{**}(t)\}}$,*
- *$q^{**}(0) = q_0$,*
- *$q^{**}(T) = 0$.*

Furthermore,

$$
\begin{aligned}
J(q^{**}) &= \int_0^T \left(V_t L\left(\frac{q^{**'}(t)}{V_t}\right) + \frac{1}{2}\gamma\sigma^2 q^{**}(t)^2 \right) dt \\
&= \int_0^T \left(V_t L\left(\frac{q^{*'}(t)}{V_t}\right) 1_{\{q^*(t) > 0, q^*(t) = q^{**}(t)\}} + \frac{1}{2}\gamma\sigma^2 q^{**}(t)^2 \right) dt.
\end{aligned}
$$

Therefore,

$$
\begin{aligned}
J(q^{**}) &\leq \int_0^T \left(V_t L\left(\frac{q^{*'}(t)}{V_t}\right) + \frac{1}{2}\gamma\sigma^2 q^{**}(t)^2 \right) dt \\
&\leq \int_0^T \left(V_t L\left(\frac{q^{*'}(t)}{V_t}\right) + \frac{1}{2}\gamma\sigma^2 q^*(t)^2 \right) dt \\
&\leq J(q^*).
\end{aligned}
$$

Using uniqueness, we have $q^* = q^{**}$.

Because q^{**} is, by construction, a nonincreasing function of time, the optimal strategy q^* is a nonincreasing function of time.

We know that the unique minimizer over $\mathcal{C}$ of the function J is a nonincreasing function if $q_0 \geq 0$ and a nondecreasing function if $q_0 \leq 0$. Therefore, q^* is also the unique minimizer of J over

$$
\mathcal{C}^- = \{q \in W^{1,1}(0, T), \operatorname{sign}(q'(t)) = -\operatorname{sign}(q_0), q(0) = q_0, q(T) = 0\}.
$$

If $L(\rho) = \tilde{L}(\rho) + \psi|\rho|$, then $\forall q \in \mathcal{C}^-$,

$$
\begin{aligned}
J(q) &= \int_0^T \left(V_t L\left(\frac{q'(t)}{V_t}\right) + \frac{1}{2}\gamma\sigma^2 q(t)^2 \right) dt \\
&= \int_0^T \left(V_t \tilde{L}\left(\frac{q'(t)}{V_t}\right) + \psi|q'(t)| + \frac{1}{2}\gamma\sigma^2 q(t)^2 \right) dt \\
&= \int_0^T \left(V_t \tilde{L}\left(\frac{q'(t)}{V_t}\right) - \psi\operatorname{sign}(q_0)q'(t) + \frac{1}{2}\gamma\sigma^2 q(t)^2 \right) dt \\
&= \int_0^T \left(V_t \tilde{L}\left(\frac{q'(t)}{V_t}\right) + \frac{1}{2}\gamma\sigma^2 q(t)^2 \right) dt + \psi|q_0| \\
&= \tilde{J}(q) + \psi|q_0|,
\end{aligned}
$$

where

$$
\tilde{J}(q) = \int_0^T \left(V_t \tilde{L}\left(\frac{q'(t)}{V_t}\right) + \frac{1}{2}\gamma\sigma^2 q(t)^2 \right) dt.
$$

Therefore, the minimizer of J over $\mathcal{C}^-$ is the minimizer of $\tilde{J}$ over $\mathcal{C}^-$. Consequently, q^* is independent of the bid-ask spread component of the execution cost function. This is an important property, but it is specific to the liquidation of a single-stock portfolio.

3.2.3.2 Characterization of the optimal strategy

To characterize the optimal strategy q^*, we can use either a Euler-Lagrange characterization or a Hamiltonian characterization. As we do not assume any regularity on the function L, the Euler-Lagrange equation is of the following form (see Theorem B.2 of Appendix B):

$$
\begin{cases}
p'(t) &= \gamma\sigma^2 q^*(t), \\
p(t) &\in \partial^- L\left(\frac{q^{*'}(t)}{V_t}\right), \\
q^*(0) &= q_0, \\
q^*(T) &= 0,
\end{cases}
$$

where we recall that $\partial^- L$ stands for the sub-differential of L. In particular, if L is differentiable, then this system reduces to

$$
\begin{cases}
p'(t) &= \gamma\sigma^2 q^*(t), \\
p(t) &= L'\left(\frac{q^{*'}(t)}{V_t}\right), \\
q^*(0) &= q_0, \\
q^*(T) &= 0.
\end{cases}
$$

This characterization is not very practical in the general case. A better characterization is with the Hamiltonian system:

$$\begin{cases} p'(t) & = & \gamma\sigma^2 q^*(t), \\ q^{*'}(t) & = & V_t H'(p(t)), \\ q^*(0) & = & q_0, \\ q^*(T) & = & 0, \end{cases} \tag{3.10}$$

where H is the Legendre-Fenchel transform of the function L defined by

$$H(p) = \sup_\rho \rho p - L(\rho).$$

Because L is strictly convex, H is a function of class C^1, and this is the reason why we can write the Hamiltonian system in the form of Eq. (3.10).

From this characterization, we can deduce that q^* is a function of class C^1 because we assume that the market volume process $(V_t)_t$ is continuous. Consequently, p is of class C^2 under the same hypothesis. However, q^* may not be of class C^2. If indeed we consider

$$L(\rho) = \eta |\rho|^{1+\phi} + \psi|\rho|,$$

then a straightforward computation gives

$$H(p) = \sup_\rho p\rho - \eta|\rho|^{1+\phi} - \psi|\rho| = \begin{cases} 0 & \text{if } |p| \le \psi, \\ \phi\eta \left(\frac{|p|-\psi}{\eta(1+\phi)} \right)^{1+\frac{1}{\phi}} & \text{otherwise.} \end{cases} \tag{3.11}$$

H is therefore C^1 but not C^2 in general, and $q^{*'}$ has no reason to be differentiable.

Another important consequence of this characterization is that the dual variable p is solution of

$$p''(t) = \gamma\sigma^2 V_t H'(p(t)) \tag{3.12}$$

with the boundary conditions $p'(0) = \gamma\sigma^2 q_0$ and $p'(T) = 0$.

It must be noticed that there may not be a unique solution to Eq. (3.12) with the above Neumann boundary conditions. For instance, if we consider H as in Eq. (3.11), and $q_0 = 0$, then any constant function p such that $\forall t \in [0, T], p(t) \in [-\psi, \psi]$ is solution of Eq. (3.12), and it satisfies the boundary conditions.

3.2.3.3   The case of quadratic execution costs

Almgren and Chriss initially introduced a model where the additional cost per share due to limited liquidity was a linear function of the number of shares.

In our framework, this corresponds to g linear, or equivalently L quadratic (see Eq. (3.3)).

To understand the model and the roles of the different parameters, let us consider a quadratic function L. If $L(\rho) = \eta \rho^2$, then, by Eq. (3.11) the associated function H is given by $H(p) = \frac{p^2}{4\eta}$. Therefore, the system (3.10) characterizing the optimal liquidation strategy becomes

$$
\begin{cases}
p'(t) &= \gamma \sigma^2 q^*(t), \\
q^{*'}(t) &= \frac{V_t}{2\eta} p(t), \\
q^*(0) &= q_0, \\
q^*(T) &= 0.
\end{cases}
\tag{3.13}
$$

Consequently, q^* is the unique solution of the equation

$$
q^{*''}(t) = \frac{\gamma \sigma^2 V_t}{2\eta} q^*(t),
$$

satisfying the boundary conditions $q^*(0) = q_0$ and $q^*(T) = 0$.

If $(V_t)_t$ is assumed to be constant (i.e., $V_t = V, \forall t \in [0, T]$), then, we get the classical hyperbolic sine formula of Almgren and Chriss:

$$
q^*(t) = q_0 \frac{\sinh\left(\sqrt{\frac{\gamma \sigma^2 V}{2\eta}}(T - t)\right)}{\sinh\left(\sqrt{\frac{\gamma \sigma^2 V}{2\eta}}T\right)}.
\tag{3.14}
$$

Associated to this optimal trading curve, the optimal (deterministic) strategy $(v_t^*)_t$ is given by

$$
v_t^* = q^{*'}(t) = -q_0 \sqrt{\frac{\gamma \sigma^2 V}{2\eta}} \frac{\cosh\left(\sqrt{\frac{\gamma \sigma^2 V}{2\eta}}(T - t)\right)}{\sinh\left(\sqrt{\frac{\gamma \sigma^2 V}{2\eta}}T\right)}.
\tag{3.15}
$$

In particular, we see from Eq. (3.14) that q^* is a convex function if $q_0 \geq 0$, and a concave function if $q_0 \leq 0$.[8] This means that the unwinding process is fast at the beginning, and decelerates progressively.

Figure 3.1 represents the optimal trading curve for a stock with the following characteristics:

- $S_0 = 45$ €,

- $\sigma = 0.6$ €·day$^{-1/2}$·share^{-1}, i.e., an annual volatility approximately equal to 21%,

---

[8]This property is true in general, under the assumption of a flat volume curve (i.e., $V_t = V, \forall t \in [0, T]$).

- $V = 4{,}000{,}000$ shares·day^{-1},

- $L(\rho) = \eta\rho^2$, with $\eta = 0.1$ €·share^{-1}.

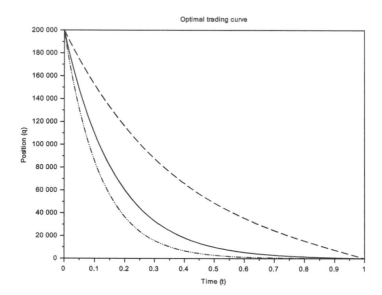

FIGURE 3.1: Optimal trading curve for $q_0 = 200{,}000$ shares over one day ($T = 1$), for different values of γ. Dash-dotted line: $\gamma = 10^{-5}$ €$^{-1}$. Solid line: $\gamma = 5.10^{-6}$ €$^{-1}$. Dashed line: $\gamma = 10^{-6}$ €$^{-1}$.

Using the closed-form formula (3.14) for the optimal trading curve, we can analyze the role of the different parameters:

The liquidity parameters η and V:

η is a scaling factor for the execution costs paid by the trader. The larger η, the more the trader pays to buy/sell shares. Similarly, the value of the market volume V is a scaling factor, because the execution costs depend on the participation rate. Hence, a small value of V has the same effect as a large value of η. In the case of a quadratic function L, we see in Eq. (3.14) that the optimal strategy only depends on the ratio $\frac{\eta}{V}$. The larger this ratio (that is, the more illiquid the stock), the slower the liquidation process:

$$\frac{d}{d\frac{\eta}{V}} \frac{q^*(t)}{q_0} \geq 0, \forall t \in [0, T].$$

The volatility parameter σ:

The volatility parameter σ measures the importance of price risk. Therefore, the larger σ, the faster the execution to reduce the exposure to price risk. This is what we observe in Eq. (3.14), as

$$\frac{d}{d\sigma} \frac{q^*(t)}{q_0} \leq 0, \forall t \in [0, T].$$

The risk aversion parameter γ:

The risk aversion parameter sets the balance between execution costs on the one hand, and price risk on the other hand. The larger the parameter γ, the more the trader is sensitive to price risk. Therefore, high γ means fast execution to reduce the exposure to price risk. This is what we observe in Eq. (3.14), as

$$\frac{d}{d\gamma} \frac{q^*(t)}{q_0} \leq 0, \forall t \in [0, T].$$

When γ is really small, we find that the optimal trading curve q^* gets close to a straight-line strategy. This is formally proved by

$$\lim_{\gamma \to 0} q^*(t) = q_0 \left(1 - \frac{t}{T} \right).$$

In contrast, the liquidation is very fast at the beginning, when γ is large. This raises the issue of imposing an upper bound to the participation rate – see Chapter 5.

3.2.4 General results

3.2.4.1 Stochastic strategies vs. deterministic strategies

So far, we have only considered the case of deterministic strategies. Our initial goal was however to obtain a control $v \in \mathcal{A}$ maximizing the objective function

$$\mathbb{E} \left[- \exp(-\gamma X_T) \right].$$

In fact, we have the following result, stating that no liquidation strategy can do better than the best deterministic strategy:

Theorem 3.2.

$$\sup_{v \in \mathcal{A}} \mathbb{E} \left[- \exp\left(-\gamma X_T \right) \right] = \sup_{v \in \mathcal{A}_{\text{det}}} \mathbb{E} \left[- \exp\left(-\gamma X_T \right) \right].$$

Proof. *It is clear that we only need to prove that*

$$\sup_{v \in \mathcal{A}} \mathbb{E} \left[- \exp\left(-\gamma X_T \right) \right] \leq \sup_{v \in \mathcal{A}_{\text{det}}} \mathbb{E} \left[- \exp\left(-\gamma X_T \right) \right].$$

To proceed, let us consider a control $v \in \mathcal{A}$. From Eq. (3.4), we know that

$$X_T = X_0 + q_0 S_0 - \frac{k}{2} q_0^2 + \sigma \int_0^T q_t dW_t - \int_0^T V_t L \left(\frac{v_t}{V_t} \right) dt.$$

Therefore,

$$\mathbb{E}\left[-\exp\left(-\gamma X_T \right) \right] = -\exp\left(-\gamma \left(X_0 + q_0 S_0 - \frac{k}{2} q_0^2 \right) \right)$$

$$\times \mathbb{E}\left[\exp\left(\gamma \left(\int_0^T V_t L \left(\frac{v_t}{V_t} \right) dt \right) \right) \exp\left(-\gamma\sigma \int_0^T q_t dW_t \right) \right].$$

Using the probability measure $\mathbb{Q}$ defined by the Radon-Nikodym derivative[9]

$$\frac{d\mathbb{Q}}{d\mathbb{P}} = \exp\left(-\gamma\sigma \int_0^T q_t dW_t - \frac{1}{2} \gamma^2 \sigma^2 \int_0^T q_t^2 dt \right),$$

we get

$$\mathbb{E}\left[-\exp\left(-\gamma X_T \right) \right] = -\exp\left(-\gamma \left(X_0 + q_0 S_0 - \frac{k}{2} q_0^2 \right) \right)$$

$$\times \mathbb{E}^{\mathbb{Q}}\left[\exp\left(\gamma \left(\int_0^T V_t L \left(\frac{v_t}{V_t} \right) dt \right) \right) \exp\left(\frac{1}{2} \gamma^2 \sigma^2 \int_0^T q_t^2 dt \right) \right].$$

Therefore,

$$\mathbb{E}\left[-\exp\left(-\gamma X_T \right) \right] = \mathbb{E}^{\mathbb{Q}}\left[-\exp\left(-\gamma \left(X_0 + q_0 S_0 - \frac{k}{2} q_0^2 - J(q) \right) \right) \right],$$

where $J(q)$ is defined almost surely, because q is almost surely in $W^{1,1}(0,T)$, by definition of $\mathcal{A}$.

Using the results of Theorem 3.1, we can write

$$\mathbb{E}\left[-\exp\left(-\gamma X_T \right) \right] \leq -\exp\left(-\gamma \left(X_0 + q_0 S_0 - \frac{k}{2} q_0^2 - J(q^*) \right) \right)$$

$$\leq \sup_{v \in \mathcal{A}_{\text{det}}} \mathbb{E}\left[-\exp\left(-\gamma X_T \right) \right].$$

By taking the supremum on the left-hand side, we get the expected result:

$$\sup_{v \in \mathcal{A}} \mathbb{E}\left[-\exp\left(-\gamma X_T \right) \right] \leq \sup_{v \in \mathcal{A}_{\text{det}}} \mathbb{E}\left[-\exp\left(-\gamma X_T \right) \right].$$

---

[9]We can apply the Girsanov theorem because $v \in \mathcal{A} \Rightarrow q$ bounded.

It is important to understand this result, both from a practical point of view and from a theoretical point of view.

For a practitioner, a deterministic strategy is a strategy that can be computed at the start of the liquidation process. In other words, whatever the evolution of the price over the course of the unwinding process, the optimal strategy, or more exactly the optimal schedule, stays the same. Consequently, the above theorem somewhat justifies the two-layer approach used in a large number of execution algorithms, where the first layer computes at time $t = 0$ the optimal schedule for the execution process – i.e., the optimal trading curve – and the second layer defines execution tactics, with all kinds of orders and on all possible venues, for liquidating the portfolio, while being as close as possible to the initial schedule.

Theorem 3.2 strongly relies on the assumptions we have made. It would not be true in general if the price was not a Brownian motion. Similarly, if the market volume process $(V_t)_t$ was not deterministic, then the above result would not hold. It is also important to note that Theorem 3.1 is specific to the CARA utility framework.

3.2.4.2 Choosing a risk profile

The expected CARA utility framework used above is equivalent to a mean-variance framework. It is mathematically convenient, but there is *a priori* no reason to think that agents have a constant absolute risk aversion.

Interestingly, Schied and Schöneborn studied in [161] the optimal liquidation of a single-stock portfolio (though only in an infinite time horizon model, i.e., $T = +\infty$), in the case of utility functions with increasing or decreasing absolute risk aversion (IARA and DARA) functions. Their result states that an agent with a IARA utility function who liquidates a long position, should accelerate the liquidation process when the price is increasing. Conversely, an agent with a DARA utility function who liquidates a long position, should decelerate the liquidation process when the price is increasing.

In practice, the choice of a utility function is always a difficult task. Even in the apparently simple case of a CARA utility function, choosing the value of the absolute risk aversion parameter γ is usually complicated – we refer to Chapters 4 and 8, where we present two different methodologies to estimate from the behavior of an agent the value he attributes to the constant γ. Furthermore, in the case of a brokerage company or a cash trader executing orders on behalf of a client, which γ should we take? The client's γ (usually unknown), or the company's/trader's γ? In the former case, the client must, in one way or another, give the value he wants for γ. In the latter case, small trades will be regarded as less risky than large trades; and this is not a good solution from a business viewpoint. Another solution is to consider for each trade a different value of γ, for instance $\gamma = \frac{\gamma'}{|q_0|S_0}$, where γ' is uniform across trades. This approach is equivalent to applying the utility function to the Profit and Loss (P&L) measured as a percentage of the initial MtM value of the portfolio (in absolute value) $|q_0|S_0$, instead of applying it to X_T, or equivalently to the P&L $X_T - X_0$.

3.3   The model in discrete time

In the previous section, we have presented a model in continuous time. Almgren and Chriss, in their first articles [8, 9], considered instead discrete-time models. Later on, in [6], Almgren introduced the continuous-time counterpart of his initial models. Both types of models are in fact interesting for different reasons. With continuous-time models, one can use differential calculus and all the tools of mathematical analysis. In contrast, discrete-time models are useful to design numerical methods in order to compute optimal trading curves in practice.

In this section, we present the discrete-time counterpart of the model presented in Section 3.2.

3.3.1   Notations

We still consider a trader with a single-stock portfolio. At time $t = 0$, his position (in number of shares) is denoted by q_0. His goal is to unwind his position by time T.

We divide the time window $[0, T]$ into N slices of length Δt. We denote by $t_0 = 0 < \ldots < t_n = n\Delta t < \ldots < t_N = N\Delta t = T$ the associated subdivision of $[0, T]$. In what follows, the index n refers either to time t_n or to the interval $[t_{n-1}, t_n]$.

At the beginning of each time interval $[t_n, t_{n+1}]$, the trader chooses the number of shares he will buy (or sell) over this time interval.[10] We denote by $v_{n+1}\Delta t$ the number of shares bought by the trader between t_n and t_{n+1} (if this number is negative, it corresponds to selling shares, instead of buying).

---

[10]As above, we are computing an optimal schedule for the execution process. We assume therefore that the trader buys (or sells) this exact number of shares.

The dynamics for the number of shares in the trader's portfolio is therefore

$$q_{n+1} = q_n + v_{n+1}\Delta t, \quad 0 \le n < N. \tag{3.16}$$

The mid-price of the stock is modeled by

$$S_{n+1} = S_n + \sigma\sqrt{\Delta t}\epsilon_{n+1} + kv_{n+1}\Delta t, \quad 0 \le n < N, \tag{3.17}$$

where $\epsilon_1, \ldots, \epsilon_N$ are independent and identically distributed (i.i.d.) $\mathcal{N}(0, 1)$ random variables.

The amount paid (received) for the $v_{n+1}\Delta t$ shares bought (sold) between t_n and t_{n+1} depends on $v_{n+1}\Delta t$, and on the market volume over $[t_n, t_{n+1}]$, denoted by $V_{n+1}\Delta t$.

The dynamics of the cash account is similar to Eq. (3.3):

$$X_{n+1} = X_n - v_{n+1}S_n\Delta t - L\left(\frac{v_{n+1}}{V_{n+1}}\right)V_{n+1}\Delta t, \quad 0 \le n < N, \tag{3.18}$$

where L verifies the same hypotheses (H1), (H2), and (H3), as in Section 3.2.1. It must be noted that the cost paid is relative to S_n, the mid-price at the beginning of the time interval $[t_n, t_{n+1}]$. In particular, we implicitly assume that there is no price risk within each time slice.

3.3.2 The optimization problem

Our goal is to find a liquidation strategy maximizing the following objective function:

$$\mathbb{E}\left[-\exp(-\gamma X_N)\right].$$

The fact that deterministic strategies are optimal can be shown in this discrete-time framework using similar techniques as in the continuous-time model. Therefore, we only consider deterministic strategies $(v_n)_n$ that are admissible, in the sense that they correspond to liquidation strategies:

$$(v_n)_n \in \mathcal{A}_d = \left\{ (v_1, \ldots, v_N) \in \mathbb{R}^N, \sum_{n=0}^{N-1} v_{n+1}\Delta t = -q_0 \right\}.$$

To solve our problem, we first compute the value of the final wealth:

$$X_N = X_0 - \sum_{n=0}^{N-1} v_{n+1}S_n\Delta t - \sum_{n=0}^{N-1} L\left(\frac{v_{n+1}}{V_{n+1}}\right)V_{n+1}\Delta t.$$

By using Eqs. (3.16), (3.17), and (3.18), we obtain

$$
\begin{aligned}
X_N &= X_0 - \sum_{n=0}^{N-1}(q_{n+1} - q_n)S_n - \sum_{n=0}^{N-1} L\left(\frac{v_{n+1}}{V_{n+1}}\right)V_{n+1}\Delta t \\
&= X_0 - \sum_{n=0}^{N-1} q_{n+1}S_n + \sum_{n=0}^{N-1} q_n S_n - \sum_{n=0}^{N-1} L\left(\frac{v_{n+1}}{V_{n+1}}\right)V_{n+1}\Delta t \\
&= X_0 - \sum_{n=0}^{N-1} q_{n+1}\left(S_{n+1} - \sigma\sqrt{\Delta t}\epsilon_{n+1} - kv_{n+1}\Delta t\right) + \sum_{n=0}^{N-1} q_n S_n \\
&\quad - \sum_{n=0}^{N-1} L\left(\frac{v_{n+1}}{V_{n+1}}\right)V_{n+1}\Delta t \\
&= X_0 + q_0 S_0 + \sigma\sqrt{\Delta t}\sum_{n=0}^{N-1} q_{n+1}\epsilon_{n+1} + k\sum_{n=0}^{N-1} q_{n+1}v_{n+1}\Delta t \\
&\quad - \sum_{n=0}^{N-1} L\left(\frac{v_{n+1}}{V_{n+1}}\right)V_{n+1}\Delta t.
\end{aligned}
$$

Let us focus on the term $k\sum_{n=0}^{N-1} q_{n+1}v_{n+1}\Delta t$. We have

$$
\begin{aligned}
k\sum_{n=0}^{N-1} q_{n+1}v_{n+1}\Delta t &= k\sum_{n=0}^{N-1} q_{n+1}(q_{n+1} - q_n) \\
&= k\sum_{n=0}^{N-1}\left(\frac{q_{n+1} + q_n}{2} + \frac{q_{n+1} - q_n}{2}\right)(q_{n+1} - q_n) \\
&= \frac{k}{2}\sum_{n=0}^{N-1}(q_{n+1}^2 - q_n^2) + \frac{k}{2}\sum_{n=0}^{N-1}(q_{n+1} - q_n)^2 \\
&= -\frac{k}{2}q_0^2 + \frac{k}{2}\sum_{n=0}^{N-1}v_{n+1}^2\Delta t^2.
\end{aligned}
$$

Therefore,

$$
\begin{aligned}
X_N &= X_0 + q_0 S_0 - \frac{k}{2}q_0^2 + \sigma\sqrt{\Delta t}\sum_{n=0}^{N-1} q_{n+1}\epsilon_{n+1} \\
&\quad + \frac{k}{2}\sum_{n=0}^{N-1}v_{n+1}^2\Delta t^2 - \sum_{n=0}^{N-1} L\left(\frac{v_{n+1}}{V_{n+1}}\right)V_{n+1}\Delta t.
\end{aligned}
$$

Because we have considered deterministic controls, the final value of the cash process X_N is normally distributed with mean

$$\mathbb{E}[X_N] = X_0 + q_0 S_0 - \frac{k}{2}q_0^2 + \frac{k}{2}\sum_{n=0}^{N-1} v_{n+1}^2 \Delta t^2 - \sum_{n=0}^{N-1} L\left(\frac{v_{n+1}}{V_{n+1}}\right) V_{n+1}\Delta t, \quad (3.19)$$

and variance

$$\mathbb{V}[X_N] = \sigma^2 \Delta t \sum_{n=0}^{N-1} q_{n+1}^2. \quad (3.20)$$

By using the same techniques as in the continuous-time case, we can compute from Eqs. (3.19) and (3.20) the value of the objective function:

$$
\begin{aligned}
\mathbb{E}\left[-\exp\left(-\gamma X_N\right)\right] &= -\exp\left(-\gamma \mathbb{E}[X_N] + \frac{1}{2}\gamma^2 \mathbb{V}[X_N]\right) \\
&= -\exp\left(-\gamma\left(X_0 + q_0 S_0 - \frac{k}{2}q_0^2\right)\right) \\
&\quad \times \exp\left(\gamma\left(\sum_{n=0}^{N-1} L\left(\frac{v_{n+1}}{V_{n+1}}\right) V_{n+1}\Delta t\right)\right) \\
&\quad \times \exp\left(\gamma\left(-\frac{k}{2}\sum_{n=0}^{N-1} v_{n+1}^2 \Delta t^2 + \frac{\gamma}{2}\sigma^2 \Delta t \sum_{n=0}^{N-1} q_{n+1}^2\right)\right).
\end{aligned}
$$

As a consequence, the problem boils down to finding a control process $(v_n)_n \in \mathcal{A}_d$ minimizing the expression

$$\sum_{n=0}^{N-1} L\left(\frac{v_{n+1}}{V_{n+1}}\right) V_{n+1}\Delta t - \frac{k}{2}\sum_{n=0}^{N-1} v_{n+1}^2 \Delta t^2 + \frac{\gamma}{2}\sigma^2 \Delta t \sum_{n=0}^{N-1} q_{n+1}^2. \quad (3.21)$$

Eq. (3.21) is the discrete-time counterpart of Eq. (3.8). An important difference is the term $\frac{k}{2}\sum_{n=0}^{N-1} v_{n+1}^2 \Delta t^2$ arising from the permanent component of market impact. If L is quadratic, then this additional term can be merged with the L term, as soon as $k\Delta t$ is small enough (to keep convexity) – see for instance [9]. If L is sub-quadratic, then this is not possible anymore. In general, because this term is a second-order term in Δt, we assume that it can be neglected.[11]

Therefore, instead of minimizing the expression in Eq. (3.21), we consider the minimization of

$$\sum_{n=0}^{N-1} L\left(\frac{v_{n+1}}{V_{n+1}}\right) V_{n+1}\Delta t + \frac{\gamma}{2}\sigma^2 \Delta t \sum_{n=0}^{N-1} q_{n+1}^2 \quad (3.22)$$

over liquidation strategies $(v_n)_n \in \mathcal{A}_d$.

---

[11] Another possibility is to replace in Eq. (3.18) the term $v_{n+1}S_n\Delta t$ by a term of the form $v_{n+1}\frac{S_n+S_{n+1}}{2}\Delta t$, but then the variance term is modified.

3.3.3   Optimal trading curve

3.3.3.1   Hamiltonian characterization

Instead of working on the liquidation strategies $(v_n)_n$, we work directly on the trading curve $(q_n)_n$. To this purpose, we introduce

$$J : q \in \mathbb{R}^{N+1} \mapsto \sum_{n=0}^{N-1} L \left(\frac{q_{n+1} - q_n}{V_{n+1} \Delta t} \right) V_{n+1} \Delta t + \frac{\gamma}{2} \sigma^2 \Delta t \sum_{n=0}^{N-1} q_{n+1}^2.$$

We want to minimize J over the set

$$\mathcal{C}_d = \{ (q_0, \ldots, q_N), q_0 = q_0, q_N = 0 \}.$$

Because J is strictly convex, and $\mathcal{C}_d$ is convex, there exists a unique minimizer of J over $\mathcal{C}_d$. We denote by q^* this unique minimizer.[12]

To characterize the optimal trading curve q^*, we use Theorem B.4 of Appendix B. We obtain that q^* is uniquely characterized by the following Hamiltonian system:

$$\begin{cases} p_{n+1} = p_n + \Delta t \gamma \sigma^2 q_{n+1}^*, & 0 \le n < N - 1, \\ q_{n+1}^* = q_n^* + \Delta t V_{n+1} H'(p_n), & 0 \le n < N, \end{cases} \qquad q_0^* = q_0, \quad q_N^* = 0.$$

$$(3.23)$$

In other words, there exists $(p_0, \ldots, p_{N-1}) \in \mathbb{R}^N$, such that (p, q^*) is a solution of Eq. (3.23), and any solution (p, q) of Eq. (3.23) is such that $q = q^*$.

This Hamiltonian system (3.23) is the discrete counterpart of the Hamiltonian system (3.10). It is a very important system of equations because many numerical methods are based on finding a solution to this system.

3.3.3.2   The initial Almgren-Chriss framework

The initial Almgren-Chriss model can be recovered from the computations we have done above, in the special case of a quadratic function L and a flat market volume curve. If $L(\rho) = \eta \rho^2$ and $\forall n, V_n = V$, then the Hamiltonian system (3.23) writes

$$\begin{cases} p_{n+1} = p_n + \Delta t \gamma \sigma^2 q_{n+1}^*, & 0 \le n < N - 1, \\ q_{n+1}^* = q_n^* + \frac{V}{2\eta} \Delta t p_n, & 0 \le n < N. \end{cases}$$

Therefore, q^* is the solution of the second-order recursive equation

$$q_{n+2}^* - \left(2 + \frac{\gamma \sigma^2 V}{2\eta} \Delta t^2 \right) q_{n+1}^* + q_n^* = 0,$$

with boundary conditions $q_0^* = q_0$ and $q_N^* = 0$.

---

[12]The optimal strategy $(v_n^*)_n \in \mathcal{A}_d$ is given by $v_n^* = \frac{q_n^* - q_{n-1}^*}{\Delta t}, 0 < n \le N$.

Solving this equation, we get

$$q_n^* = q_0 \frac{\sinh\left(\alpha(T - t_n)\right)}{\sinh\left(\alpha T\right)},$$

where α the unique positive solution of

$$2\left(\cosh(\alpha \Delta t) - 1\right) = \frac{\gamma \sigma^2 V}{2\eta} \Delta t^2.$$

In particular, when $\Delta t \to 0$, we recover Eq. (3.14).

Furthermore, it must be noted that, in the case of a quadratic function L given by $L(\rho) = \eta \rho^2$, and when $\forall n, V_n = V$, Eq. (3.21) becomes

$$\sum_{n=0}^{N-1} \eta \frac{v_{n+1}^2}{V} \Delta t - \frac{k}{2} \sum_{n=0}^{N-1} v_{n+1}^2 \Delta t^2 + \frac{\gamma}{2} \sigma^2 \Delta t \sum_{n=0}^{N-1} q_{n+1}^2$$

$$= \sum_{n=0}^{N-1} \left(\frac{\eta - \frac{k}{2} V \Delta t}{V}\right) v_{n+1}^2 \Delta t + \frac{\gamma}{2} \sigma^2 \Delta t \sum_{n=0}^{N-1} q_{n+1}^2.$$

Therefore, if $\eta > \frac{k}{2} V \Delta t$, then we can replace η by $\tilde{\eta} = \eta - \frac{k}{2} V \Delta t$ if we want to take into account the second-order term coming from permanent market impact.

3.4 Conclusion

In this chapter, we have introduced the basic model for the optimal scheduling of an execution process. The limited available liquidity is modeled with a general strictly convex execution cost function. The permanent component of market impact is modeled with a linear function, that guarantees the absence of dynamic arbitrage. The model we have presented, which is a generalization of the basic Almgren-Chriss model, leads to an optimal trading curve that can be used in the strategic layer of execution algorithms. However, several questions arise from this model.

First, we have only dealt with single-stock portfolios. Chapter 5 will cover the liquidation of a complex portfolio, and also generalize the model of this chapter to include some additional features.

Second, we have characterized the optimal trading curve with the Hamiltonian system (3.10), or its discrete counterpart Eq. (3.23). However, we have not dealt with the numerical methods to approximate the solutions of these equations. What would be the most rapid or the most general one? This is the purpose of Chapter 6.

Third, we have only dealt with optimal scheduling, that is, the first layer of execution algorithms. What about the second layer? This is the purpose of Chapter 7.

Furthermore, the optimal execution models presented in this chapter only focus on one kind of execution: the execution of IS (Implementation Shortfall) orders. This means that the price risk is measured with respect to the initial/arrival price S_0. In the academic literature, optimal execution models most often deal, explicitly or implicitly, with these IS orders. However, other orders are used by practitioners, and are in fact used far more often than IS orders. The goal of the next chapter is to discuss the main types of orders used in the brokerage industry, and to show how the models presented in this chapter can be adapted, in order to optimally schedule the execution process of many types of orders.

Chapter 4

Optimal liquidation with different benchmarks

Par ma foi ! Il y a plus de quarante ans que je dis de la prose sans
que j'en susse rien, et je vous suis le plus obligé du monde de m'avoir
appris cela.

— Monsieur Jourdain, *in* Le Bourgeois Gentilhomme, Molière

Like Monsieur Jourdain, the character of Molière who has been speaking
prose for years without even knowing it, most academic papers on optimal
execution deal with a specific type of orders, often without knowing it: Implementation Shortfall (IS) orders. In this chapter, we use the Almgren-Chriss
framework introduced in Chapter 3 to deal with other types of orders: Target Close, POV, and VWAP orders.[1] We also discuss the limitations of the
Almgren-Chriss framework, especially as far as market volumes are concerned.

4.1 Introduction: the different types of orders

The academic literature on optimal execution most often deals with the optimal liquidation of a portfolio – most often a single-stock one. We have seen
in the previous chapter what optimality could mean: maximizing a mean-
variance objective function[2] where the main variable is the P&L of the liquidation strategy. Other objective functions are sometimes used in the academic
literature, but the mathematical problem almost always boils down to maximizing a risk-adjusted function of the P&L. This approach is natural, but it
corresponds to only one of the numerous types of orders proposed by brokers:
Implementation Shortfall (IS) orders.

In practice, brokers and cash trading desks of investment banks propose a
large list of services to their clients to buy or sell stocks. In addition to providing direct market access (DMA), they usually propose a swath of strategies

---

[1]It must be noted that POV or VWAP orders are used more often than IS orders in
practice.

[2]In fact an expected utility objective function that boils down to a mean-variance one.

to execute orders on behalf of their clients.[3] These strategies may be different from one broker to another, but there are usually at least five types of strategies/orders that are common to all of them – sometimes with different names:

- Implementation Shortfall (IS) orders,

- Target Close (TC) orders,

- Percentage Of Volume (POV) orders,

- Volume-Weighted Average Price (VWAP) orders,

- Time-Weighted Average Price (TWAP) orders.

IS orders correspond to the strategy discussed in the previous chapter. The trading curve is optimized to minimize execution costs while mitigating price risk, the risk being measured with respect to the price at the start. IS orders are also called arrival price orders; the goal is to obtain a price that is not too different from the market price at the start of the execution process, while minimizing execution costs. IS orders are for instance used by market participants to benefit from a price opportunity.

Contrary to IS orders, for which the benchmark is the arrival price, TC orders correspond to a strategy benchmarked to the closing price of the day. If there is an auction at the end of the day, orders can be executed at this price. However, large orders sent during the closing auction may have a significant impact on the closing price, and it is often preferable to execute part of large orders beforehand, during the continuous auction period. If there is no closing auction, and if the closing price is fixed by another mechanism,[4] then, whatever the size of the order, a strategy is required to obtain a price that is not too far from the closing price. By using TC orders, agents aim at minimizing execution costs, while mitigating the risk of being executed at a price too different from the closing price. TC orders are for instance used by the managers of funds for which the NAV (Net Asset Value) is computed using closing prices. An important difference with IS orders is that the benchmark price is unknown at the start of the execution process.

POV orders correspond to a strategy in which the volume executed over the course of the process must be as close as possible to a fixed constant percentage of the market volume – this percentage being decided upon at the start of the execution process, usually by the broker's client. These orders are used by market participants for executing large blocks of shares while following the market flow.

---

[3] Here, we only consider agency trades, for which the client of the broker bears the risk of the execution. We shall discuss in Chapter 8 the case of principal trades (risk trades).

[4] For instance in Hong Kong – see Chapter 2.

VWAP and TWAP orders correspond to strategies benchmarked to the average price of the stock over a period of time chosen at the beginning of the execution process. In the case of VWAP orders, the average price is volume-weighted: it is the ratio of the total value (volume × price) of all trades over the period to the total volume transacted over the same period. In the case of TWAP orders, it is the time-weighted average price. These types of orders are usually chosen to buy or sell a medium-to-large quantity of shares over a given period of time without much execution costs. They follow the market flow, but in a different way than POV orders. In the former case, the execution period is fixed at the start of the execution process, while in the latter case the participation rate to the market is fixed and the execution period depends therefore on market volumes. As for TC orders, the benchmark price is unknown at the beginning of the execution process.

In this chapter, we first show that TC orders are very similar to IS orders as far as mathematical modeling is concerned. In particular, we show how the trading curve of a TC order can be obtained from the trading curve of an IS order. With respect to POV orders, we derive the optimal participation rate as a function of liquidity, volatility, and risk aversion (see also [81]). The closed-form formula we obtain for the optimal participation rate may also serve as a way to estimate risk aversion. VWAP orders are then considered in the last section of this chapter, with both models *à la* Almgren-Chriss and other types of models.

4.2 Target Close orders

4.2.1 Target Close orders in the Almgren-Chriss framework

In Europe, depending on the day, closing auctions can represent from 5% up to 20% of the total daily volume. Closing prices play indeed a major role as the computation of the NAV of most funds relies on them. This is one of the reasons why many agents want to be executed at a price that is as close as possible to the closing price. However, very large orders are seldom sent to be executed at the closing auction, for fear that they contribute too much to the closing price. A basic strategy consists instead in (i) deciding[5] which quantity is going to be executed during the closing auction, and (ii) executing the rest of the order during the continuous auction, before the closing auction.

Target Close orders are used by market participants to obtain a price as close as possible to the closing price of a trading day, while trading during the

---

[5]The decision is usually based upon an estimate of the total volume executed in the closing auction.

continuous auction. In terms of mathematical modeling, we keep the framework introduced by Almgren-Chriss. We consider a trader with a single-stock portfolio, and an initial position denoted by q_0 (positive or negative, depending on whether the trader is long or short). We consider the problem of unwinding the portfolio by the end of the day. We assume that the trader can trade during the continuous auction and at the closing auction. To simplify the model, we assume that the trader chooses in advance a (signed) quantity v_{close} to send at the closing auction.[6] The trader's position over the time interval $[0, T]$ – corresponding to the continuous auction – is modeled by the process $(q_t)_{t \in [0,T]}$. The dynamics of this process is

$$dq_t = v_t dt, \tag{4.1}$$

where $(v_t)_{t \in [0,T]}$ is a control in the set $\mathcal{A}_{\text{close}}$ of admissible controls defined by

$$\mathcal{A}_{\text{close}} = \left\{ (v_t)_{t \in [0,T]} \in \mathbb{H}^0(\mathbb{R}, (\mathcal{F}_t)_t), \right.$$

$$\left. \int_0^T v_t dt + v_{\text{close}} = -q_0, \int_0^T |v_t| dt \in L^\infty(\Omega) \right\}.$$

The mid-price of the stock during the continuous auction is modeled by the process $(S_t)_{t \in [0,T]}$ following the dynamics

$$dS_t = \sigma dW_t + k v_t dt. \tag{4.2}$$

As far as execution costs are concerned, we consider an execution cost function L similar to those of Chapter 3. Therefore, during the continuous trading phase, the cash account process $(X_t)_{t \in [0,T]}$ evolves as

$$dX_t = -v_t S_t dt - V_t L \left(\frac{v_t}{V_t} \right) dt, \tag{4.3}$$

where $(V_t)_t$ is the market volume process, assumed to be deterministic, continuous, positive, and bounded.

To model the closing auction, we simply assume that the closing price is

$$S_{\text{close}} = S_T + h(v_{\text{close}}) + \sigma_{\text{close}} \epsilon, \tag{4.4}$$

where ϵ is a random variable, normally distributed $\mathcal{N}(0, 1)$, independent of $(W_t)_t$, and where h is an impact function. For the problem under consideration, it is not necessary to distinguish which proportion of the price impact at the closing auction will permanently impact the stock price, and which proportion will only be temporary.[7]

---

[6] Obviously we assume that $v_{\text{close}} \in [0, -q_0)$, for the problem to be meaningful.

[7] The absence of dynamic arbitrage in this context is guaranteed as soon as $kqv + \frac{k}{2}v^2 \leq h(v)(q + v), \forall (q, v) \in \mathbb{R}^2$.

The cash account at the end of the closing auction is

$$X_{\text{close}} = X_T - v_{\text{close}} S_{\text{close}}. \tag{4.5}$$

The problem we consider is the maximization of the objective function

$$\mathbb{E}\left[-\exp(-\gamma(X_{\text{close}} - X_0 - q_0 S_{\text{close}}))\right],$$

over admissible controls $(v_t)_{t \in [0,T]} \in \mathcal{A}_{\text{close}}$.

In other words, the utility function applies to the slippage between the P&L and the benchmark P&L, which is the initial position q_0 multiplied by the closing price S_{close}, and not to the P&L,[8] as in the case of IS orders (see Chapter 3).

4.2.2 Target Close orders as reversed IS orders

To solve the above problem, we first compute the slippage

$$X_{\text{close}} - X_0 - q_0 S_{\text{close}}.$$

By using Eq. (4.5), we have

$$X_{\text{close}} - X_0 - q_0 S_{\text{close}} \;=\; X_T - X_0 - (v_{\text{close}} + q_0) S_{\text{close}}.$$

By using Eqs. (4.1), (4.2), (4.3), and (4.4) and an integration by parts, we get

$$X_{\text{close}} - X_0 - q_0 S_{\text{close}}$$

$$= -\int_0^T v_t S_t dt - \int_0^T V_t L\left(\frac{v_t}{V_t}\right) dt - (v_{\text{close}} + q_0) S_{\text{close}}$$

$$= -(v_{\text{close}} + q_0)(S_{\text{close}} - S_T) - \int_0^T k v_t (q_0 - q_t) dt$$

$$-\sigma \int_0^T (q_0 - q_t) dW_t - \int_0^T V_t L\left(\frac{v_t}{V_t}\right) dt$$

$$= -(v_{\text{close}} + q_0) h(v_{\text{close}}) - (v_{\text{close}} + q_0)\sigma_{\text{close}}\epsilon + \frac{k}{2}(q_0 + v_{\text{close}})^2$$

$$-\sigma \int_0^T (q_0 - q_t) dW_t - \int_0^T V_t L\left(\frac{v_t}{V_t}\right) dt.$$

---

[8]It is noteworthy that, in the case of IS orders, the utility function could be applied equivalently to the cash account at the end of the process, the P&L, or the slippage between the P&L and $q_0 S_0$, which is the benchmark P&L in the case of an IS order.

Therefore, using the independence between ϵ and $(W_t)_t$, and the Laplace transform of a Gaussian variable (see Theorem A.1 of Appendix A), the objective function writes

$$\mathbb{E}\left[-\exp(-\gamma(X_{\text{close}} - X_0 - q_0 S_{\text{close}}))\right]$$

$$= \exp\left(-\gamma\left(-(v_{\text{close}} + q_0)h(v_{\text{close}}) + \frac{k}{2}(q_0 + v_{\text{close}})^2\right)\right)$$

$$\times \exp\left(\frac{1}{2}\gamma^2\sigma_{\text{close}}^2(v_{\text{close}} + q_0)^2\right)$$

$$\times \mathbb{E}\left[-\exp\left(-\gamma\left(-\sigma\int_0^T (q_0 - q_t)dW_t - \int_0^T V_t L\left(\frac{v_t}{V_t}\right)dt\right)\right)\right].$$

As a consequence, the problem boils down to maximizing

$$\mathbb{E}\left[-\exp\left(-\gamma\left(-\sigma\int_0^T (q_0 - q_t)dW_t - \int_0^T V_t L\left(\frac{v_t}{V_t}\right)dt\right)\right)\right]$$

over the set $\mathcal{A}_{\text{close}}$ of admissible controls.

Using the same reasoning as in Chapter 3, we can search for an optimal control in the subset of $\mathcal{A}_{\text{close}}$ consisting of deterministic controls only. If v_t is deterministic, then $t \mapsto q(t) = q_t$ is an absolutely continuous function, and we have

$$\mathbb{E}\left[-\exp\left(-\gamma\left(-\sigma\int_0^T (q_0 - q_t)dW_t - \int_0^T V_t L\left(\frac{v_t}{V_t}\right)dt\right)\right)\right]$$

$$= -\exp\left(\gamma\left(\int_0^T \left(V_t L\left(\frac{q'(t)}{V_t}\right) + \frac{1}{2}\gamma\sigma^2(q_0 - q(t))^2\right)dt\right)\right).$$

Therefore, the problem is a variational problem that consists in minimizing the function

$$J_{TC}(q) = \int_0^T \left(V_t L\left(\frac{q'(t)}{V_t}\right) + \frac{1}{2}\gamma\sigma^2(q_0 - q(t))^2\right)dt,$$

over the set of functions $q \in W^{1,1}(0,T)$ satisfying the constraints $q(0) = q_0$ and $q(T) = -v_{\text{close}}$.

From Theorem B.3 of Appendix B, we know that this problem has a solution. Moreover, the solution is unique because of the strict convexity of J_{TC}.

In fact, the solution of this problem can be linked to the solution of the classical problem of an IS order dealt with in Chapter 3. To highlight this point, we introduce a new function $\tilde{q}(t) = q_0 - q(T - t)$ and a new process $\tilde{V}_t = V_{T-t}$. We see that

$$
\begin{aligned}
J_{TC}(q) &= \int_0^T \left(V_{T-t} L \left(\frac{q'(T-t)}{V_{T-t}} \right) + \frac{1}{2} \gamma \sigma^2 (q_0 - q(T-t))^2 \right) dt \\
&= \int_0^T \left(V_{T-t} L \left(\frac{\tilde{q}'(t)}{V_{T-t}} \right) + \frac{1}{2} \gamma \sigma^2 \tilde{q}(t)^2 \right) dt \\
&= \int_0^T \left(\tilde{V}_t L \left(\frac{\tilde{q}'(t)}{\tilde{V}_t} \right) + \frac{1}{2} \gamma \sigma^2 \tilde{q}(t)^2 \right) dt \\
&= \tilde{J}(\tilde{q}),
\end{aligned}
$$

where

$$
\tilde{J}(\tilde{q}) = \int_0^T \left(\tilde{V}_t L \left(\frac{\tilde{q}'(t)}{\tilde{V}_t} \right) + \frac{1}{2} \gamma \sigma^2 \tilde{q}(t)^2 \right) dt.
$$

Therefore, minimizing J_{TC} over the set of functions $q \in W^{1,1}(0, T)$ satisfying the constraints $q(0) = q_0$ and $q(T) = -v_{\text{close}}$ is equivalent to minimizing $\tilde{J}$ over the set of functions $\tilde{q} \in W^{1,1}(0, T)$ satisfying the constraints $\tilde{q}(0) = q_0 + v_{\text{close}}$ and $\tilde{q}(T) = 0$.

We can state this result in the form of a theorem:[9]

Theorem 4.1. *Let* $\tilde{V}_t = V_{T-t}$, *and let*

$$
\tilde{J} : \tilde{q} \in W^{1,1}(0, T) \mapsto \int_0^T \left(\tilde{V}_t L \left(\frac{\tilde{q}'(t)}{\tilde{V}_t} \right) + \frac{1}{2} \gamma \sigma^2 \tilde{q}(t)^2 \right) dt.
$$

The objective function

$$
(v_t)_t \in \mathcal{A}_{close} \mapsto \mathbb{E} \left[-\exp(-\gamma(X_{close} - X_0 - q_0 S_{close})) \right],
$$

is maximized when $v_t = \tilde{q}^{*'}(T-t)$, *where* $\tilde{q}^*$ *is the unique minimizer of* $\tilde{J}$ *over the set of functions* $\tilde{q} \in W^{1,1}(0, T)$ *satisfying the constraints* $\tilde{q}(0) = q_0 + v_{close}$ *and* $\tilde{q}(T) = 0$.

The most remarkable point is that the minimization of $\tilde{J}$ is exactly the problem of finding the optimal trading curve of an IS order (see Chapter 3) to unwind a position $q_0 + v_{\text{close}}$ in the stock, when the market volume curve $(V_t)_t$ is replaced by the time-reversed one $(\tilde{V}_t)_t$. In other words, a Target Close order is nothing but a reversed IS order on the part of the order that is not executed during the closing auction.

---

[9]The optimality of deterministic strategies can be proven using the same trick as in Theorem 3.2.

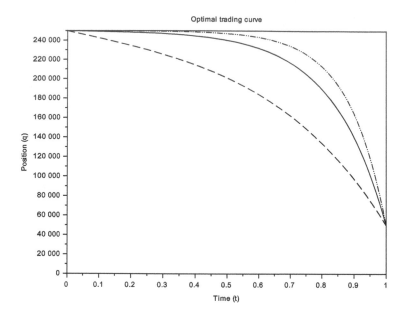

FIGURE 4.1: Optimal trading curves for a Target Close order for $q_0 = $ 250,000 shares over one day $(T = 1)$, when $v_{\text{close}} = -50,000$ shares, for different values of γ. Dash-dotted line: $\gamma = 10^{-5}$ €$^{-1}$. Solid line: $\gamma = 5.10^{-6}$ €$^{-1}$. Dashed line: $\gamma = 10^{-6}$ €$^{-1}$.

To illustrate our results, we have plotted in Figure 4.1 the optimal trading curve associated with a Target Close order to sell 250,000 shares, when 50,000 shares are executed at the closing auction, for a stock with the following characteristics:

- $S_0 = 45$ €,

- $\sigma = 0.6$ €·day$^{-1/2}$·share^{-1}, i.e., an annual volatility approximately equal to 21%,

- $V = 4{,}000{,}000$ shares·day^{-1},

- $L(\rho) = \eta\rho^2$ with $\eta = 0.1$ €·share^{-1}.

4.2.3   Concluding remarks on Target Close orders

The Almgren-Chriss framework makes it possible to build a very simple model for finding the optimal trading curve associated with a Target Close order. Assuming that the quantity to be executed at the closing auction is decided upon *ex-ante*, the problem can be transformed to be similar to the one tackled in Chapter 3. In particular, a TC order can be seen as a form of reversed IS order.

This result is interesting, but several questions remain unanswered. How should one choose v_{close}? Could not we imagine a model in which the quantity to be executed at the closing auction evolves over the course of the execution process?

The choice of v_{close} raises in fact deep questions about Target Close orders. If the true goal of an agent in his choice to use a Target Close order was to be executed as close as possible to the closing price, then it would be optimal for him to execute the whole quantity at the closing auction. In practice, this is obviously nonsense for large orders. In fact, the goal of an agent when he chooses to use a Target Close order is to be executed as close as possible to the closing price, without contributing too much to its value. Consequently, a way to make the choice of v_{close} endogenous is to replace the benchmark price S_{close} by another price, such as a convex combination between S_T and S_{close}, and to solve the model by using techniques of variational calculus. However, in practice, choosing the weights of the convex combination is as difficult as choosing v_{close} directly. In practice, when v_{close} is chosen prior to the execution process, it is often chosen based on an estimate of the future market volume at the closing auction.

The second question is deeper, and it highlights the limitations of the Almgren-Chriss framework. Estimates of the future market volume at the closing auction can be computed prior to the start of the execution process, but also over the course of the execution process. It is always possible to compute a new trading curve at any time to take account of new information, but this raises a problem of time inconsistency. The solution is in fact to consider a different stochastic optimal control problem with an additional state variable related to an estimate of the market volume at the closing auction. But then, the problem may be difficult to solve, even numerically, because there are already three state variables (cash, position, and price). This problem goes in fact beyond the case of TC orders; some variables, such as the market volume, would indeed better be modeled with stochastic processes than with deterministic processes.

4.3   POV orders

4.3.1   Presentation of the problem

As above, we consider the case of a trader with a single-stock portfolio containing q_0 shares. Here, we assume that the trader is long, i.e., $q_0 > 0$,[10] and we assume that he is willing to unwind his portfolio. Contrary to the above cases however, we do not specify a time window for the execution process. Instead, we consider that the trader chooses to liquidate at a constant participation rate to the market, denoted by ρ.[11]

The position of the trader is modeled by a process $(q_t)_{t\in\mathbb{R}_+}$ defined on $\mathbb{R}^+$. Its dynamics is

$$dq_t = v_t dt, \tag{4.6}$$

with $(v_t)_t$ in the set $\mathcal{A}_{POV}$ of admissible controls defined here by

$$\mathcal{A}_{POV} = \left\{ (v_t)_{t\in\mathbb{R}_+}, \exists \rho > 0, \forall t \geq 0, v_t = -\rho V_t 1_{\{\int_0^t \rho V_s ds \leq q_0\}} \right\},$$

where $(V_t)_t$ is the market volume process, assumed to be deterministic, continuous, and such that $0 < \underline{V} \leq V_t \leq \overline{V}$.

The process $(S_t)_t$ for the mid-price of the stock is of the same form as before:

$$dS_t = \sigma dW_t + k v_t dt. \tag{4.7}$$

As far as execution costs are concerned, we consider asymptotically superlinear execution cost functions L similar to those of Chapter 3. Therefore, the cash account process $(X_t)_t$ evolves as

$$dX_t = -v_t S_t dt - V_t L\left(\frac{v_t}{V_t}\right) dt. \tag{4.8}$$

Our goal is to maximize over $(v_t)_t \in \mathcal{A}_{POV}$ the objective criterion

$$\mathbb{E}\left[-\exp(-\gamma X_T)\right],$$

where T is such that $\int_0^T v_t dt = -q_0$ and $\gamma > 0$ is, as always, the absolute risk aversion parameter of the trader.[12]

---

[10]The case $q_0 < 0$ can be treated in a similar way.

[11]In practice, for agency trades, the client chooses the rate of participation – often a round figure. For risk trades, however, the trader/broker may be able to choose any participation rate.

[12]Another objective function that leads to the same results is

$$\mathbb{E}\left[-\exp(-\gamma(X_T - X_0 - q_0 S_0))\right].$$

Contrary to the case of IS (or TC) orders, in the case of a POV strategy, the volume traded is constrained to be proportional to the market volume process. The problem we consider here is therefore the choice[13] of the participation rate ρ.[14]

4.3.2 Optimal participation rate

The above optimization problem is not a complex variational problem because there is only one parameter to optimize upon: the participation rate ρ.

Let us compute the value X_T of the cash account process at the end of the execution process when $v_t = -\rho V_t$ over the course of the execution process. By using Eqs. (4.6), (4.7), and (4.8) and an integration by parts, we get

$$
\begin{aligned}
X_T &= X_0 - \int_0^T v_t S_t dt - \int_0^T V_t L\left(\frac{v_t}{V_t}\right) dt \\
&= X_0 + q_0 S_0 + \int_0^T k v_t q_t dt + \int_0^T q_t \sigma dW_t - \int_0^T V_t L\left(\rho\right) dt \\
&= X_0 + q_0 S_0 - \frac{k}{2} q_0^2 + \sigma \int_0^T \left(q_0 - \rho \int_0^t V_s ds\right) dW_t - \frac{L(\rho)}{\rho} q_0 \\
&= X_0 + q_0 S_0 - \frac{k}{2} q_0^2 + \sigma \rho \int_0^T \int_t^T V_s ds dW_t - \frac{L(\rho)}{\rho} q_0.
\end{aligned}
\tag{4.9}
$$

From Eq. (4.9), we deduce that the final value of the cash account process X_T is normally distributed with mean

$$
\mathbb{E}[X_T] = X_0 + q_0 S_0 - \frac{k}{2} q_0^2 - \frac{L(\rho)}{\rho} q_0,
\tag{4.10}
$$

and variance

$$
\mathbb{V}[X_T] = \sigma^2 \rho^2 \int_0^T \left(\int_t^T V_s ds\right)^2 dt.
\tag{4.11}
$$

By using the Laplace transform of a Gaussian variable (see Theorem A.1 of Appendix A), and Eqs. (4.10) and (4.11), we have

$$
\mathbb{E}\left[-\exp\left(-\gamma X_T\right)\right] = \mathbb{E}\left[-\exp\left(-\gamma \mathbb{E}[X_T] + \frac{\gamma^2}{2} \mathbb{V}[X_T]\right)\right].
$$

---

[13]It is noteworthy that, in practice, for agency trades, this choice is made by the client and not by the broker or the cash trader.

[14]The real participation rate is $\frac{\rho}{1+\rho}$ because of the trader's own volume. However, we call ρ the participation rate throughout this section.

As a consequence, we see that maximizing

$$\mathbb{E}\left[-\exp\left(-\gamma X_T\right)\right] \;=\; -\exp\left(-\gamma\left(X_0 + q_0 S_0 - \frac{k}{2}q_0^2\right)\right)$$

$$\times \exp\left(\gamma\left(\frac{L(\rho)}{\rho}q_0 + \frac{\gamma}{2}\sigma^2\rho^2\int_0^T\left(\int_t^T V_s ds\right)^2 dt\right)\right)$$

boils down to finding a minimum to the function

$$J_{\text{POV}} \;:\; \mathbb{R}_+^* \;\to\; \mathbb{R}$$

$$\rho \;\mapsto\; \frac{L(\rho)}{\rho}q_0 + \frac{\gamma}{2}\sigma^2\rho^2\int_0^T\left(\int_t^T V_s ds\right)^2 dt,$$

where T satisfies $\rho\int_0^T V_s ds = q_0$.

We can now prove that J_{POV} has a global minimum ρ^*, but the most important result of this section is the expression of ρ^*, in the particular case of an execution cost function L given by $L(\rho) = \eta|\rho|^{1+\phi}$ and a flat market volume curve.

Theorem 4.2. *There exists $\rho^* > 0$ such that J_{POV} has a global minimum in ρ^*.*

Furthermore, if the execution function L is given by $L(\rho) = \eta|\rho|^{1+\phi}$, with $\phi > 0$, and the market volume is assumed to be constant, with $V_t = V$, then ρ^* is unique, and given by

$$\rho^* = \left(\frac{\gamma\sigma^2}{6\eta\phi}\frac{q_0^2}{V}\right)^{\frac{1}{1+\phi}}. \tag{4.12}$$

Proof. *Because J_{POV} is continuous, in order to prove the existence of a global minimum in the general case, we can just prove that*

$$\lim_{\rho\to 0} J_{POV}(\rho) = \lim_{\rho\to+\infty} J_{POV}(\rho) = +\infty.$$

Because L is assumed to be asymptotically super-linear, we have $\lim_{\rho\to+\infty}\frac{L(\rho)}{\rho}q_0 = +\infty$ and therefore $\lim_{\rho\to+\infty} J_{POV}(\rho) = +\infty$.

Now, using the bounds on $(V_t)_t$, we have

$$\int_0^T\left(\int_t^T V_s ds\right)^2 dt \geq \underline{V}^2\frac{T^3}{3},$$

and

$$q_0 = \rho\int_0^T V_t dt \leq \rho\overline{V}T.$$

Therefore,

$$\rho^2 \int_0^T \left(\int_t^T V_s ds \right)^2 dt \geq \frac{1}{3} \rho^2 \underline{V}^2 T^3 \geq \frac{1}{3} \underline{V}^2 \frac{q_0^3}{\overline{V}^3} \frac{1}{\rho},$$

and we obtain $\lim_{\rho \to 0} J_{POV}(\rho) = +\infty.$

We conclude that there exists indeed a minimizer $\rho^* \in \mathbb{R}_+^*$ *of* $J_{POV}.$

In the special case of a power function $\rho \in \mathbb{R}^+ \mapsto L(\rho) = \eta |\rho|^{1+\phi} = \eta \rho^{1+\phi}$ *and a flat market volume curve* $V_t = V$, *we have*

$$J_{POV}(\rho) = \eta \rho^\phi q_0 + \frac{\gamma}{2} \sigma^2 \rho^2 V^2 \frac{T^3}{3}.$$

Because $q_0 = \rho V T$, *we have*

$$J_{POV}(\rho) = \eta \rho^\phi q_0 + \frac{\gamma}{6} \sigma^2 \frac{q_0^3}{\rho V}.$$

Therefore,

$$J'_{POV}(\rho) = \eta \phi \rho^{\phi-1} q_0 - \frac{\gamma}{6} \sigma^2 \frac{q_0^3}{\rho^2 V},$$

and

$$J'_{POV}(\rho) = 0 \iff \eta \phi \rho^{\phi-1} = \frac{\gamma}{6} \sigma^2 \frac{q_0^2}{\rho^2 V} \iff \rho = \left(\frac{\gamma \sigma^2}{6 \eta \phi} \frac{q_0^2}{V} \right)^{\frac{1}{1+\phi}}.$$

We conclude that ρ^* *is unique, with*

$$\rho^* = \left(\frac{\gamma \sigma^2}{6 \eta \phi} \frac{q_0^2}{V} \right)^{\frac{1}{1+\phi}}.$$

The closed-form formula (4.12) for the optimal participation rate deserves several comments about the influence of the parameters:

- ρ^* is an increasing function of γ and σ. The higher the risk aversion parameter γ or the volatility σ, the faster the execution in order to reduce price risk.

- ρ^* is an increasing function of q_0. This is related to the fact that the larger his inventory, the more a trader is exposed to price risk, and therefore the faster the execution.

- The instantaneous volume executed by the trader, i.e., $\rho^* V$, is given by $\rho^* V = \left(\frac{\gamma \sigma^2}{6 \eta} q_0^2 \right)^{\frac{1}{1+\phi}} V^{\frac{\phi}{1+\phi}}$. It is an increasing function of V and a decreasing function of η. This means that the more liquid the stock, the more volume we trade per unit of time.

In order to illustrate our results, let us consider the same example as before, where

- $S_0 = 45$ €,

- $\sigma = 0.6$ €·day$^{-1/2}$·share^{-1}, i.e., an annual volatility approximately equal to 21%,

- $V = 4,000,000$ shares·day^{-1},

- $L(\rho) = \eta\rho^2$ with $\eta = 0.1$ €·share^{-1},

- $q_0 = 200,000$ shares.

Table 4.1 gives the optimal participation rate for several values of the risk aversion parameter γ. We see that the optimal participation rate depends heavily on the risk aversion parameter γ.

TABLE 4.1: Optimal participation rate ρ^* for different values of the risk aversion parameter γ.

Risk aversion parameter γ (€$^{-1}$)	Optimal participation rate ρ^*
10^{-7}	6.7%
5.10^{-7}	11.4%
10^{-6}	14.4%
5.10^{-6}	24.7%
10^{-5}	31.1%

4.3.3   A way to estimate risk aversion

In practice, Eq. (4.12) provides a way to be consistent across stocks and across market conditions, thanks to a common parameter γ. Most of the papers in the literature introduce one or several parameters playing a similar role as our risk aversion parameter γ. However, most authors remain silent on the estimation of these parameters; this is a problem, given the influence of γ (see Table 4.1).

We believe that Eq. (4.12) may be very useful both to estimate γ and to be coherent across execution processes. Eq. (4.12) can indeed be rewritten as

$$\gamma = 6\frac{\eta V \phi}{\sigma^2}\frac{\rho^{*\phi+1}}{q_0^2}.$$

This new equation could be used to derive the risk aversion parameter of a market participant from the observation of his behavior, or from his answers to questions about the rhythm at which he would optimally unwind different portfolios in different market conditions.

4.4 VWAP orders

VWAP and TWAP orders are often used by market practitioners as a way to obtain a price as close as possible to the average price over a given period of time. In this chapter, we focus on VWAP orders, but the case of TWAP orders can be treated in the same way, given that the two benchmark prices (the TWAP and the VWAP) coincide in the case of a flat market volume curve.

The VWAP benchmark is a natural benchmark for utilitarian traders who do not send orders in order to benefit from a price opportunity, but instead to buy or sell shares in line with their global investment strategies or to hedge a risky position. These traders want to have as little execution costs as possible, and want to pay a price that corresponds to the average price paid over the period to buy or sell the stock.

Most of the literature on optimal execution is about IS orders, but there exist a few papers on VWAP orders. In this section, we first present the optimal strategy associated with a VWAP order in the Almgren-Chriss framework. Then, we highlight the limitations of the Almgren-Chriss framework, and we discuss the alternative models proposed in the academic and professional literature.

4.4.1 VWAP orders in the Almgren-Chriss framework

4.4.1.1 The model

We consider a trader with a single-stock portfolio. As, always, his position at time $t = 0$ is denoted by q_0. The problem is to unwind the portfolio over a time window $[0, T]$.

The trader's position over the time interval $[0, T]$ is modeled by the process $(q_t)_{t \in [0,T]}$. The dynamics of this process is

$$dq_t = v_t dt, \tag{4.13}$$

where $(v_t)_{t \in [0,T]}$ is in the set of admissible processes $\mathcal{A}$ defined by

$$\mathcal{A} = \left\{ (v_t)_{t \in [0,T]} \in \mathbb{H}^0(\mathbb{R}, (\mathcal{F}_t)_t), \int_0^T v_t dt = -q_0, \int_0^T |v_t| dt \in L^\infty(\Omega) \right\}.$$

In addition to the trader's volume, we introduce the market volume process $(V_t)_t$, assumed to be deterministic, continuous, positive, and bounded.

The cumulated market volume from 0 up to time t is denoted by Q_t:

$$Q_t = \int_0^t V_s ds.$$

The mid-price of the stock is modeled by the process $(S_t)_{t\in[0,T]}$. As above, we assume that the dynamics of the mid-price is of the form

$$dS_t = \sigma dW_t + kv_t dt. \tag{4.14}$$

Eventually, the cash account process $(X_t)_{t\in[0,T]}$ has the following dynamics:

$$dX_t = -v_t S_t dt - V_t L\left(\frac{v_t}{V_t}\right) dt, \tag{4.15}$$

where L is an execution cost function satisfying the same properties as in Chapter 3.

This framework is exactly the same as for an IS order. However, in the case of a VWAP order over the period $[0,T]$, the benchmark price VWAP_T is given by[15]

$$\mathrm{VWAP}_T = \frac{\int_0^T S_t V_t dt}{\int_0^T V_t dt} = \frac{\int_0^T S_t dQ_t}{Q_T}. \tag{4.16}$$

Therefore, the problem is to maximize

$$\mathbb{E}\left[-\exp\left(-\gamma(X_T - X_0 - q_0 \mathrm{VWAP}_T)\right)\right],$$

over $(v_t)_{t\in[0,T]} \in \mathcal{A}$.

For that purpose, let us compute the slippage $X_T - X_0 - q_0 \mathrm{VWAP}_T$. By using Eqs. (4.13), (4.14), and (4.15), the definition (4.16) of the VWAP, and several integrations by parts, we get

$$X_T - X_0 - q_0 \mathrm{VWAP}_T$$

$$= -\int_0^T v_t S_t dt - \int_0^T V_t L\left(\frac{v_t}{V_t}\right) dt - q_0 \frac{\int_0^T S_t dQ_t}{Q_T}$$

$$= q_0 S_0 + \int_0^T k v_t q_t dt + \sigma \int_0^T q_t dW_t - \int_0^T V_t L\left(\frac{v_t}{V_t}\right) dt$$

$$- \frac{q_0}{Q_T}\left(Q_T S_0 + \sigma \int_0^T (Q_T - Q_t) dW_t + \int_0^T k v_t (Q_T - Q_t) dt\right).$$

---

[15]In this chapter, we consider the case of a small order and we neglect our own volume. In general, the VWAP would rather be defined by

$$\mathrm{VWAP}_T = \frac{\int_0^T S_t(V_t + |v_t|) dt}{\int_0^T (V_t + |v_t|) dt}.$$

Therefore,

$$
X_T - X_0 - q_0 \text{VWAP}_T = -\frac{k}{2}q_0^2 + \sigma \int_0^T \left(q_t - q_0 \left(1 - \frac{Q_t}{Q_T} \right) \right) dW_t
$$
$$
- \int_0^T V_t L \left(\frac{v_t}{V_t} \right) dt + k q_0 \int_0^T \frac{V_t}{Q_T}(q_0 - q_t) dt.
$$

Using the same reasoning as in Chapter 3, we can show that no control in $v \in \mathcal{A}$ can do better than the best deterministic control in $\mathcal{A}$. Therefore, we consider deterministic strategies only. In that case, using the Laplace transform of a Gaussian variable (see Theorem A.1 of Appendix A), the objective function writes

$$
\mathbb{E}\left[-\exp(-\gamma(X_T - X_0 - q_0 \text{VWAP}_T)) \right]
$$
$$
= -\exp\left(-\gamma \left(-\frac{k}{2}q_0^2 - \int_0^T V_t L \left(\frac{v_t}{V_t} \right) dt + k q_0 \int_0^T \frac{V_t}{Q_T}(q_0 - q_t) dt \right) \right)
$$
$$
\times \exp\left(\frac{1}{2}\gamma^2 \left(q_t - q_0 \left(1 - \frac{Q_t}{Q_T} \right) \right)^2 \right).
$$

Therefore, the problem boils down to minimizing the function J_{VWAP} defined by
$$
J_{\text{VWAP}}(q) =
$$
$$
\int_0^T \left(V_t L \left(\frac{q'(t)}{V_t} \right) - k q_0 \frac{V_t}{Q_T}(q_0 - q(t)) + \frac{1}{2}\gamma\sigma^2 \left(q(t) - q_0 \left(1 - \frac{Q_t}{Q_T} \right) \right)^2 \right) dt,
$$

over the set of functions $q \in W^{1,1}(0,T)$ satisfying the constraints $q(0) = q_0$ and $q(T) = 0$.

This new objective function deserves a few comments, because it is not as simple as in the case of IS or TC orders. The permanent market impact appears indeed in the objective function, and the optimal strategy will therefore depend on k. In the case of an IS order as in the case of a VWAP order, it is straightforward to verify that the additional cost paid by the trader because of the permanent market is the same: $\frac{k}{2}q_0^2$. When it comes to the benchmark price, the situation is different. In the case of an IS order, the benchmark price S_0 does not depend on k. In the case of a VWAP order, the benchmark price is

$$
\text{VWAP}_T = \frac{\int_0^T S_t dQ_t}{Q_T} = S_0 + \sigma \int_0^T \left(1 - \frac{Q_t}{Q_T} \right) dW_t + \int_0^T k v_t \left(1 - \frac{Q_t}{Q_T} \right) dt.
$$

Therefore, in the case of a VWAP order, the benchmark price is impacted, not only by the total executed quantity, but also by the execution strategy itself. In particular, in the case of a VWAP sell order, the optimal strategy consists in selling more at the start, in order to push down the price at the beginning of the process. Then, the VWAP computed over the period $[0, T]$ will be lower than it would have been without the excess selling at the beginning.

To better understand the problem, let us first consider the case without permanent market impact. If $k = 0$, then the problem can be solved in closed-form.

Theorem 4.3. *If $k = 0$, then the unique minimizer of J_{VWAP} over the set of functions $q \in W^{1,1}(0, T)$ satisfying the constraints $q(0) = q_0$ and $q(T) = 0$ is*

$$q^*(t) = q_0 \left(1 - \frac{Q_t}{Q_T}\right).$$

Proof. *By definition of J_{VWAP}, for all $q \in W^{1,1}(0, T)$ satisfying the constraints $q(0) = q_0$ and $q(T) = 0$, we have, by Jensen's inequality, that*

$$
\begin{aligned}
J_{VWAP}(q) &= \int_0^T \left(V_t L\left(\frac{q'(t)}{V_t}\right) + \frac{1}{2}\gamma\sigma^2\left(q(t) - q_0\left(1 - \frac{Q_t}{Q_T}\right)\right)^2\right) dt \\
&\geq \int_0^T V_t L\left(\frac{q'(t)}{V_t}\right) dt \\
&\geq Q_T L\left(\int_0^T \frac{q'(t)}{V_t}\frac{V_t}{Q_T} dt\right) \\
&= Q_T L\left(-\frac{q_0}{Q_T}\right) \\
&= J_{VWAP}(q^*).
\end{aligned}
$$

Therefore q^* minimizes J_{VWAP}, and it is the unique minimizer because J_{VWAP} is strictly convex.

As suggested by intuition, when there is no permanent market impact, Theorem 4.3 states that the optimal strategy associated with a VWAP order consists in selling at a constant participation rate, in order to be executed at the VWAP (minus execution costs). It is noteworthy that this result still holds when the definition of the VWAP includes our own trades.

In the general case, when permanent market impact is taken into account, the minimization of the function J_{VWAP} turns out to be a search for a balance between three effects:

- impacting the market price to push up/down the VWAP (depending on whether the order is a buy or sell order),[16]

---

[16]This remark raises questions about whether or not considering VWAP as a benchmark

- minimizing the costs related to the execution cost,

- minimizing the deviation from the naïve strategy $q(t) = q_0 \left(1 - \frac{Q_t}{Q_T}\right)$, which is optimal when permanent market impact is not taken into account.

There is no closed-form solution in the general case.[17] However, we can prove that there exists a unique solution, and provide a differential characterization of the optimal solution.

Theorem 4.4. *There exists a unique minimizer of J_{VWAP} over the set of functions $q \in W^{1,1}(0,T)$ satisfying the constraints $q(0) = q_0$ and $q(T) = 0$. It is uniquely characterized by the following Hamiltonian system:*

$$
\begin{cases}
p'(t) & = & \gamma\sigma^2 \left(q^*(t) - q_0\left(1 - \frac{Q_t}{Q_T}\right)\right) + kq_0\frac{V_t}{Q_T}, \\
q^{*'}(t) & = & V_t H'(p(t)), \\
q^*(0) & = & q_0, \\
q^*(T) & = & 0.
\end{cases}
\tag{4.17}
$$

Proof. *The existence of a minimizer is obtained with Theorem B.3 of Appendix B. The uniqueness comes from the strict convexity of J_{VWAP}. The Hamiltonian characterization of Eq. (4.17) is a direct application of Theorem B.2 of Appendix B.*

Theorem 4.4 proves that there is a unique optimum in the class of deterministic admissible controls. The same trick as in Theorem 3.2 applies to prove the general optimality of the optimal deterministic strategy.

4.4.1.2 Examples and analysis

So far, we have presented optimal trading curves in the case of a flat market volume curve. In the case of a VWAP order, the main feature of the strategy is that it follows the market volume curve,[18] and only deviates from it when there is permanent market impact.

In Figure 4.2, we see the market volume curve of the stock BNP Paribas over one day (June 15, 2015). This market volume curve is obviously unknown prior to the execution process, and a model is needed to estimate the market volume curve.

---

is a good option. However, the size of the order has to be large and the permanent market impact high, for the impact on the strategy to be significant.

[17]In the case of quadratic execution costs with a flat market volume curve, there is a closed-form formula. We refer the interested reader to [91].

[18]In the case of POV orders, the optimal strategy also follows the market volume curve. However, the goal of the model for POV orders is to determine the optimal participation rate. Therefore, considering a flat market volume curve as an approximation does not have severe consequences on the output of the model for POV orders.

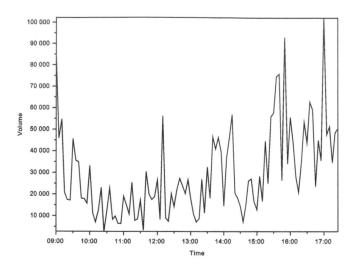

FIGURE 4.2: Market volume curve for the stock BNP Paribas on June 15, 2015 on Euronext (5-minute bins). Auctions are not represented.

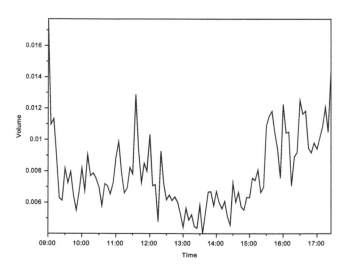

FIGURE 4.3: Average relative market volume curve for the stock BNP Paribas in June 2015 (special days have been removed) on Euronext (5-minute bins). Auctions are not represented.

Although the total daily market volume is often difficult to predict, the shape of the market volume curve is usually consistent across (normal) days. In Europe for instance (see Figure 4.3),[19] we see, on average, a decreasing level of activity over the first part of the day. Then, we see a progressive increase in market volumes that starts around the opening hour of the New York markets and continues until the closing auction, because many actors want to be executed at a price close to the closing price. All over the world, relative market volume curves[20] $t \mapsto \frac{V_t}{Q_T}$ are often U-shaped with more activity at the beginning and at the end of the day.

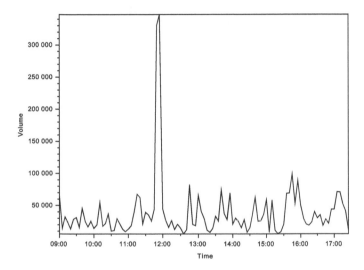

FIGURE 4.4: Market volume curve (no auction) for the stock BNP Paribas on June 19, 2015 on Euronext (5-minute bins). The date corresponds to the third Friday of June: a settlement day for derivative products on several Stoxx indices – the settlement price being the average of the index values calculated between 11:50 and 12:00 CET.

Furthermore, it is important to take account of special days, in which the market volume curve is impacted by specific events: the release of important macroeconomic statistics – that occurs on a monthly or quarterly basis – the expiration of futures and option contracts on major indices (witching days), etc. For instance, in Figure 4.4, we have plotted the market volume curve for the stock BNP Paribas on June 19, 2015 on Euronext. The date corre-

---

[19]In practice, relative market volume curves are often smoothed, using for instance cubic splines or any well-chosen basis of polynomials.

[20]Sometimes, the graph of $t \mapsto \frac{Q_t}{Q_T}$ is also called the relative volume curve.

sponds to the third Friday of June, which is both the third Friday of a month and the third Friday of a quarter-end month. It is therefore the settlement day for many derivative products on several Stoxx indices (for instance the very important Stoxx 50 Index whose constituents include BNP Paribas). In particular, the settlement price for many of these derivative products is the average of the values of the underlying index calculated between 11:50 and 12:00 CET. This explains why market volumes skyrocket just before noon, and why the shape of the market volume curve is so particular on that day.

Figure 4.5 shows the trading curve associated with a VWAP order for $q_0 = 200,000$ shares over one day, when permanent market impact is not taken into account, and when the relative market volume curve considered to schedule execution is the one of Figure 4.3. The trading curve is not a straight line: execution accelerates when the market volume is high and decelerates in periods of low market volume (especially between 12:30 and 14:30).

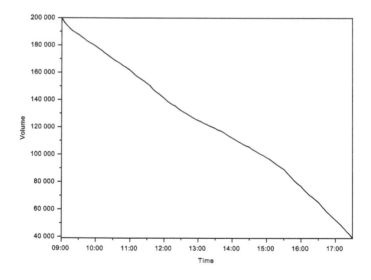

FIGURE 4.5: Optimal trading curves for a VWAP order for $q_0 = 200,000$ shares over one day.

The Almgren-Chriss framework enables us to deal with VWAP orders, but the model is based on a pre-computed market volume curve. The optimal strategy for an IS order or a TC order is also based on a pre-computed market volume curve, but the changes in the market volume only change the magnitude of execution costs. In the case of a VWAP order, the benchmark price mainly depends on the market volume curve, and the optimal strategy consists in following the volume flow at the right pace, computed *ex-ante*. Therefore, in terms of trading cost analysis (TCA), an error in the estimation of the market volume curve is far more important for a VWAP order than for any other type of orders.[21] Furthermore, the relation between market volumes and volatility is absent from the Almgren-Chriss framework, even though it can be introduced in a weak form, through the replacement of the constant volatility parameter σ, by a deterministic process $(\sigma_t)_t$ – see also Chapter 5.

4.4.2 Other models for VWAP orders

Notwithstanding the importance of VWAP orders in algorithmic trading, both conceptually and in terms of the overall volume traded using VWAP algorithms, there are only a few papers about VWAP orders in the academic literature. Furthermore, most of these papers have been written with little reference to the literature on optimal execution.

The first important paper tackling execution with VWAP benchmark is a paper published in 2002 by Konishi [117]. Konishi developed a model in which volatility and market volumes are stochastic (contrary to the case we presented above), but he did not consider any form of execution costs or market impact. His goal was to find the deterministic (or static) trading curve, that minimizes the variance of the slippage.

With our notations, Konishi's model writes

$$
\begin{aligned}
dS_t &= \sigma_t dW_t, \\
dX_t &= -S_t dq(t), \\
\text{VWAP}_T &= \frac{1}{Q_T} \int_0^T S_t dQ_t,
\end{aligned}
$$

where $(\sigma_t)_t$ is a positive stochastic process.

Then, the problem is to find $t \mapsto q(t)$, with $q(0) = q_0$ and $q(T) = 0$, that minimizes

$$
\mathbb{V}(X_T - X_0 - q_0 \text{VWAP}_T).
$$

In what follows, we do not detail the mathematical hypotheses for Konishi's results to hold. We simply present the main ideas put forward by Konishi in his paper.

---

[21] Many papers study and model the dynamics and seasonality of intraday market volumes. The interested readers should for instance read the paper of Bialkowski et al. [22], with application to VWAP trading, or the paper of McCulloch [141].

His approach is based on the following computations:

$$\mathbb{V}(X_T - X_0 - q_0 \text{VWAP}_T)$$

$$= \mathbb{V}\left(-\int_0^T S_t dq(t) - \frac{q_0}{Q_T}\int_0^T S_t dQ_t\right)$$

$$= \mathbb{V}\left(\int_0^T \left(q(t) - q_0\left(1 - \frac{Q_t}{Q_T}\right)\right)\sigma_t dW_t\right)$$

$$= \int_0^T \mathbb{E}\left[\sigma_t^2\left(q(t) - q_0\left(1 - \frac{Q_t}{Q_T}\right)\right)^2\right] dt$$

$$= \int_0^T \left(q(t)^2 \mathbb{E}\left[\sigma_t^2\right] - 2q(t)q_0\mathbb{E}\left[\sigma_t^2\left(1 - \frac{Q_t}{Q_T}\right)\right] + q_0^2\mathbb{E}\left[\sigma_t^2\left(1 - \frac{Q_t}{Q_T}\right)^2\right]\right) dt.$$

Therefore, the optimal trading curve is given by

$$q^*(t) = q_0 \frac{\mathbb{E}\left[\sigma_t^2\left(1 - \frac{Q_t}{Q_T}\right)\right]}{\mathbb{E}\left[\sigma_t^2\right]}.$$

This optimal strategy can be rewritten as

$$q^*(t) = q_0\mathbb{E}\left[1 - \frac{Q_t}{Q_T}\right] - q_0\frac{\text{Cov}\left(\sigma_t^2, \frac{Q_t}{Q_T}\right)}{\mathbb{E}\left[\sigma_t^2\right]}. \tag{4.18}$$

Konishi's model is interesting in that it is completely orthogonal to the Almgren-Chriss framework. There is indeed no market impact, but stochastic market volumes and stochastic volatility.

In Konishi's model, when the volatility process is independent from (or even uncorrelated with) the market volume process, the optimal strategy consists in following the expected trajectory of the relative market volume curve:

$$q^*(t) = q_0\mathbb{E}\left[1 - \frac{Q_t}{Q_T}\right].$$

However, when the volatility process is correlated with the market volume process, the optimal strategy consists in a deviation from the expected trajectory of the relative market volume curve, the deviation depending on the covariance between the square of the volatility process and the relative market volume curve process. This is Eq. (4.18). In practice, increases in volume are positively correlated with increases in volatility. Therefore, in the case of a sell order, the trading curve in the model of Konishi departs from the expected trajectory of the relative market volume curve, and is below it.

Konishi's model was extended by McCulloch and Kazakov [142]. In particular, they considered a mean-variance objective function, in order to take into account a drift in the price process. They also considered the addition of participation constraints. Moreover, they discussed a discrete-time version of their model.

The model we have built using the Almgren-Chriss framework and Konishi's model focus on different aspects of optimal execution with VWAP benchmark. However, they are both leading to deterministic trading curves. Dynamic models have also been proposed. For instance, McCulloch and Kazakov proposed another paper [143] in which the optimal trajectory is obtained by projecting the solution when the total volume Q_T is known onto the space of $(\mathcal{F}_t)_t$-adapted processes. Li also proposed a more advanced execution model with both market orders and limit orders, and a VWAP benchmark price – see [132, 133]. We shall not comment on all the papers on VWAP trading,[22] but the paper [73] by Frei and Westray deserves a few words. It is indeed one of the only papers (with [39] and [91]) that tackles optimal execution with VWAP benchmark in a framework with (quadratic) execution costs. This paper, based on a classical approach with stochastic optimal control and Hamilton-Jacobi-Bellman equations, is particularly interesting because Frei and Westray provide a result on the unique way to model (by using a Gamma bridge) the market volume curve $(Q_t)_t$ under the assumption that $\left(\frac{Q_t}{Q_T}\right)_t$ is independent from Q_T. For more details on this interesting approach, we refer the reader to [73].

4.5 Conclusion

In this chapter, we have seen how the Almgren-Chriss framework could be used to deal with the main types of orders used in practice. We hope to have convinced the reader that this framework is a very useful one to optimally schedule execution. In all cases, the models we have built indeed provide a trading curve that can be used in the first (strategic) layer of execution algorithms.

In the case of IS and TC orders (and also in the case of VWAP orders, when permanent market impact is taken into account), the optimal trading curve is the minimizer of a convex variational problem, characterized by a Hamiltonian system. The numerical resolution of the Hamiltonian systems

---

[22]Other papers include [28, 39, 112, 114, 146, 175].

involved in optimal execution is one of the topics of Chapter 6. Before moving to numerical methods, Chapter 5 focuses on several generalizations of the models introduced in Chapters 3 and 4. In particular, the case of multi-stock portfolios is addressed, for tackling the issues faced by quantitative analysts in Program Trading (PT).

The orders considered in this chapter are the main types of agency orders proposed by brokers and cash traders to their clients. In the case of agency orders, the risk is borne by the client. This is not the case for principal trades (or risk trades). In Chapter 8, we use the models introduced in Chapter 3 and Chapter 4, to price risk trades, i.e., to price blocks of shares, or to price Guaranteed VWAP contracts.

Chapter 5

Extensions of the Almgren-Chriss framework

There is a joke that your hammer will always find nails to hit. I find that perfectly acceptable.

— Benoît Mandelbrot

In this short chapter, we present several extensions of the Almgren-Chriss framework. In the first section, we focus on some very basic generalizations of the model. We show that the trader can incorporate his views on the future behavior of the market price of the stock. In particular, we show that a drift can very easily be added. Furthermore, we show that the volatility parameter σ can easily be replaced by a volatility process $(\sigma_t)_t$, as long as this process is deterministic. In the second section, we show how participation constraints can be added, for avoiding trading too fast. In the third section, we generalize the Almgren-Chriss framework to the case of a multi-stock portfolio. We cover the case of IS orders, but TC or VWAP orders can be tackled in the same way.

5.1   A more complex price dynamics

The first extension we consider is related to the dynamics of the stock price. We consider the introduction of a deterministic drift into the price dynamics, which can reflect the views of the trader on the future dynamics of the market price. Unwinding a long position when prices have a negative drift is different from unwinding the same portfolio when prices have a positive drift. In the former case, it is indeed more urgent to liquidate than in the latter. This extension can be particularly useful in the case of a liquidation under stress, or in the case of risk trades, when the bank suspects some form of toxicity. For some specific clients, it makes indeed sense to assume that there will be an upward trend following their decision to buy, and a downward trend following their decision to sell. In addition to a drift, we consider a nonconstant volatility process, which reflects the intraday seasonality of volatility.[1]

5.1.1   The model

We consider a model that is very similar to the continuous-time model of Chapter 3. A trader with a single-stock portfolio, and an initial position denoted by q_0, is willing to unwind his portfolio over a time window $[0, T]$. The trader's position over the time interval $[0, T]$ is modeled by the process $(q_t)_{t \in [0,T]}$. Its dynamics is

$$dq_t = v_t dt,$$

where $(v_t)_{t \in [0,T]}$ belongs to the set of admissible controls $\mathcal{A}$ given by

$$\mathcal{A} = \left\{ (v_t)_{t \in [0,T]} \in \mathbb{H}^0(\mathbb{R}, (\mathcal{F}_t)_t), \int_0^T v_t dt = -q_0, \int_0^T |v_t| dt \in L^\infty(\Omega) \right\}.$$

The difference with the model of Chapter 3 is that the process $(S_t)_{t \in [0,T]}$ for the mid-price of the stock has the following dynamics:

$$dS_t = \mu_t dt + \sigma_t dW_t + k v_t dt,$$

where $(\mu_t)_t$ and $(\sigma_t)_t$ are two deterministic processes.[2]

The cash account process $(X_t)_{t \in [0,T]}$ evolves as

$$dX_t \;=\; -v_t S_t dt - V_t L\left(\frac{v_t}{V_t}\right) dt,$$

where L satisfies the usual assumptions, and $(V_t)_{t \in [0,T]}$ is a deterministic, continuous, positive, and bounded process.

The goal of this section is to show how to find a maximizer of

$$\mathbb{E}\left[-\exp(-\gamma X_T) \right],$$

over the set of admissible controls $\mathcal{A}$, in this generalized model.

5.1.2   Extension of the Hamiltonian system

As in Chapter 3, the first step of the reasoning consists in computing X_T.

---

[1] Compared with the intraday seasonality of market volumes, this is a second-order effect for execution problems.
[2] $(\sigma_t)_t$ is assumed to be a positive process.

In the present model, we have

$$
\begin{aligned}
X_T &= X_0 - \int_0^T v_t S_t dt - \int_0^T V_t L\left(\frac{v_t}{V_t}\right) dt \\
&= X_0 + q_0 S_0 + \int_0^T k v_t q_t dt + \int_0^T q_t \mu_t dt \\
&\quad + \int_0^T q_t \sigma_t dW_t - \int_0^T V_t L\left(\frac{v_t}{V_t}\right) dt \\
&= X_0 + q_0 S_0 - \frac{k}{2} q_0^2 + \int_0^T q_t \mu_t dt \\
&\quad + \int_0^T q_t \sigma_t dW_t - \int_0^T V_t L\left(\frac{v_t}{V_t}\right) dt.
\end{aligned}
$$

The same reasoning as in Chapter 3 applies, and the problem boils down to finding a deterministic control process $(v_t)_{t\in[0,T]} \in \mathcal{A}$ minimizing

$$
\int_0^T V_t L\left(\frac{v_t}{V_t}\right) dt - \int_0^T \mu_t q_t dt + \frac{\gamma}{2} \int_0^T \sigma_t^2 q_t^2 dt.
$$

This problem can be written as a Bolza problem, where one has to find the minimizer of the function J defined by

$$
J : q \in \mathcal{C} \mapsto \int_0^T \left(V_t L\left(\frac{q'(t)}{V_t}\right) - \mu_t q(t) + \frac{1}{2}\gamma \sigma_t^2 q(t)^2 \right) dt,
$$

where

$$
\mathcal{C} = \left\{ q \in W^{1,1}(0,T), q(0) = q_0, q(T) = 0 \right\}.
$$

By using Theorems B.2 and B.3 of Appendix B, and the same reasoning as in Chapter 3, we can state the following result:

Theorem 5.1. *There exists a unique minimizer q^* of the function J over the set $\mathcal{C}$.*

Furthermore, q^* is uniquely characterized by

$$
\begin{cases}
p'(t) &= -\mu_t + \gamma \sigma_t^2 q^*(t), \\
q^{*'}(t) &= V_t H'(p(t)), \\
q^*(0) &= q_0, \\
q^*(T) &= 0,
\end{cases}
\tag{5.1}
$$

where H is the Legendre-Fenchel transform of the function L.

Theorem 5.1 shows that generalizing the model of Chapter 3 by adding a drift and a time dependence on the volatility parameter is very easy (as long as the volatility process is not stochastic). However, the monotonicity property of trading curves is lost in general; it may be optimal to do a round trip to benefit from the drift.

In this extension, it must be noted that two theoretical problems are in fact entangled: a liquidation problem and an asset management problem. If there was no liquidation problem, the optimal portfolio choice in this model would classically be $q_{am}(t) = \frac{\mu_t}{\gamma \sigma_t^2}$. Therefore, the Hamiltonian system (5.1) can also be written as

$$
\begin{cases}
p'(t) & = & \gamma \sigma_t^2 \left(q^*(t) - q_{am}(t) \right), \\
q^{*\,'}(t) & = & V_t H'(p(t)), \\
q^*(0) & = & q_0, \\
q^*(T) & = & 0.
\end{cases}
$$

In particular, when $(\mu_t)_t$ and $(\sigma_t)_t$ are constant processes, $q_{am} = \frac{\mu}{\gamma \sigma^2}$ is a constant and the optimal trading curve is given by

$$
q^*(t) = \frac{\mu}{\gamma \sigma^2} + q(t),
$$

where $t \mapsto q(t)$ is the unique solution of the Hamiltonian system

$$
\begin{cases}
p'(t) & = & \gamma \sigma^2 q(t), \\
q'(t) & = & V_t H'(p(t)), \\
q(0) & = & q_0 - \frac{\mu}{\gamma \sigma^2}, \\
q(T) & = & -\frac{\mu}{\gamma \sigma^2}.
\end{cases}
$$

It is noteworthy that this Hamiltonian system does not correspond to a portfolio liquidation problem, but instead to a portfolio transition one.

5.2   Adding participation constraints

In the model of Chapter 3, the execution cost function prevented the trader from unwinding his position too fast. This soft constraint on the participation rate is however too soft in some cases, and it may be interesting to add a harder constraint on the participation rate. Constraining the participation rate to stay below a certain threshold may also be a wise decision, because the execution cost functions we have used in Chapters 3 and 4 are often calibrated on trades of small-to-medium size, and therefore inaccurate for high participation rates (say above 20%). In this section, we cover the case of IS orders with a maximum threshold on the participation rate.

5.2.1 The model

As above, we consider the case of a trader with a single-stock portfolio, and an initial position denoted by q_0. We consider the classical problem of unwinding this portfolio over a time window $[0, T]$. The difference with the model of Chapter 3 is that the trader's position over the time interval $[0, T]$ is modeled by the process $(q_t)_{t \in [0,T]}$, which has the following dynamics:

$$dq_t = v_t dt,$$

where $(v_t)_{t \in [0,T]}$ belongs to the set of admissible controls $\mathcal{A}_{\rho_{\max}}$ defined by

$$\mathcal{A}_{\rho_{\max}} = \left\{ (v_t)_{t \in [0,T]} \in \mathbb{H}^0(\mathbb{R}, (\mathcal{F}_t)_t), \int_0^T v_t dt = -q_0, |v_t| \le \rho_{\max} V_t, \text{a.e.} \right\},$$

where $\rho_{\max} > 0$ is the upper bound for the participation rate, and $(V_t)_{t \in [0,T]}$ is the market volume process that represents the volume traded by other agents. As above, we assume that $(V_t)_{t \in [0,T]}$ is a deterministic, continuous, positive, and bounded process. To ensure that $\mathcal{A}_{\rho_{\max}} \ne \emptyset$, we also assume that

$$|q_0| \le \rho_{\max} \int_0^T V_t dt.$$

The process $(S_t)_{t \in [0,T]}$ for the mid-price of the stock has the classical dynamics:

$$dS_t = \sigma dW_t + k v_t dt.$$

The cash account process $(X_t)_{t \in [0,T]}$ evolves as

$$dX_t = -v_t S_t dt - V_t L \left(\frac{v_t}{V_t} \right) dt,$$

where L satisfies the usual assumptions.

The goal of this section is to show how to find a maximizer of

$$\mathbb{E} \left[- \exp(-\gamma X_T) \right],$$

over the new set of admissible controls $\mathcal{A}_{\rho_{\max}}$.

5.2.2 Towards a new Hamiltonian system

The same reasoning as in Chapter 3 applies (see mainly Section 3.2.3). We have

$$X_T = X_0 + q_0 S_0 - \frac{k}{2} q_0^2 + \sigma \int_0^T q_t dW_t - \int_0^T V_t L \left(\frac{v_t}{V_t} \right) dt,$$

and the problem boils down to finding a deterministic control process $(v_t)_{t\in[0,T]} \in \mathcal{A}_{\rho\max}$ minimizing

$$\int_0^T V_t L\left(\frac{v_t}{V_t}\right) dt + \frac{\gamma}{2}\sigma^2 \int_0^T q_t^2 dt.$$

This problem can be written as a Bolza problem with a generalized functional J defined by

$$J : q \in \mathcal{C} \mapsto \int_0^T \left(V_t L_{\rho\max}\left(\frac{q'(t)}{V_t}\right) + \frac{1}{2}\gamma\sigma^2 q(t)^2\right) dt,$$

where

$$\mathcal{C} = \left\{q \in W^{1,1}(0,T), q(0) = q_0, q(T) = 0\right\},$$

and

$$L_{\rho\max} : \rho \in \mathbb{R} \mapsto \begin{cases} L(\rho) & \text{if } |\rho| \leq \rho_{\max}, \\ +\infty & \text{otherwise.} \end{cases}$$

The main trick here is to put the constraint into the objective function, not into the set of admissible functions.

By using Theorems B.2 and B.3 of Appendix B, and the same reasoning as in Chapter 3, we can state the following result:

Theorem 5.2. *There exists a unique minimizer q^* of the function J over the set $\mathcal{C}$.*

If $q_0 \geq 0$, then q^* is a nonincreasing function of time. If $q_0 \leq 0$, then q^* is a nondecreasing function of time.

Furthermore, q^* is uniquely characterized by

$$\begin{cases} p'(t) &=& \gamma\sigma^2 q^*(t), \\ q^{*'}(t) &=& V_t H'_{\rho\max}(p(t)), \\ q^*(0) &=& q_0, \\ q^*(T) &=& 0, \end{cases} \tag{5.2}$$

where $H_{\rho\max}$ is the Legendre-Fenchel transform of the function $L_{\rho\max}$ defined by

$$H_{\rho\max}(p) = \sup_{|\rho|\leq\rho_{\max}} \rho p - L(\rho).$$

Theorem 5.2 states that adding a participation constraint simply boils down to changing the Hamiltonian system from Eq. (3.10) to Eq. (5.2) – the only change is in the Hamiltonian function H that becomes $H_{\rho\max}$.

In the special case of a function L of the form

$$L(\rho) = \eta |\rho|^{1+\phi} + \psi |\rho|, \qquad (\phi > 0)$$

the Hamiltonian function is

$$H_{\rho_{\max}}(p) = \begin{cases} 0 & \text{if } |p| \le \psi, \\ \phi \eta \left(\frac{|p| - \psi}{\eta(1+\phi)} \right)^{1+\frac{1}{\phi}} & \text{if } \psi < |p| \le \psi + \eta(1+\phi)\rho_{\max}^{\phi}, \\ (|p| - \psi)\rho_{\max} - \eta \rho_{\max}^{1+\phi} & \text{if } \psi + \eta(1+\phi)\rho_{\max}^{\phi} < |p|. \end{cases}$$

$H'_{\rho_{\max}}$ is therefore given by

$$H'_{\rho_{\max}}(p) = \text{sign}(p) \min \left(\rho_{\max}, \left(\frac{\max(|p| - \psi, 0)}{\eta(1+\phi)} \right)^{\frac{1}{\phi}} \right).$$

We will see, in Chapter 6, several methods to approximate the solutions of the Hamiltonian systems of both Eq. (3.10) and Eq. (5.2).

5.2.3 What about a minimal participation rate?

So far, we have only discussed the addition of an upper bound to the participation rate. Sometimes, it may be interesting to impose a minimum participation rate. It is indeed nonsense to trade a very small number of shares.

This problem may look similar to the previous one, but it is not. Requiring a minimum participation rate means in fact that the trader can choose to trade or not, and, only if he trades, then the volume must be such that the participation rate is above the minimum threshold. In mathematical terms, the problem is therefore not convex. It consists in minimizing a function of the form

$$J : q \in \mathcal{C}_{\rho_{\min}} \mapsto \int_0^T \left(V_t L \left(\frac{q'(t)}{V_t} \right) + \frac{1}{2}\gamma\sigma^2 q(t)^2 \right) dt,$$

where

$$\mathcal{C}_{\rho_{\min}} = \left\{ q \in W^{1,1}(0,T), q(0) = q_0, q(T) = 0, \right.$$

$$\left. |q'(t)| \ge \rho_{\min} V_t \text{ a.e. on } \{q'(t) \ne 0\} \right\},$$

and $\mathcal{C}_{\rho_{\min}}$ is not a convex set.

In practice, the problem can be solved if we only consider convex (respectively concave) trading curves for the liquidation of a long (respectively short) position.[3] The idea is to impose to trade at each period (with the constraint of a minimum participation rate) until a time $\tau \leq T$ and then to stop trading. For a fixed τ, the problem is convex.[4] Then, the problem boils down to numerically finding the best τ.

This question of a minimum participation rate is therefore tractable in practice – see for instance [122] – but only in the case of a single-stock portfolio. In the multi-asset case, the non-convexity of the problem makes it very complex.

5.3   Portfolio liquidation

So far, we have only tackled optimal execution problems in the case of single-stock portfolios. In this section, we generalize the Almgren-Chriss model to the case of multi-stock portfolios.

This generalization is important, for instance, for Program Trading (PT). Program Trading is often used by funds to buy or sell several stocks over a given period of time, in order to rebalance their positions, or after an increase or decrease in AuM (Assets under Management). Buying and selling several shares at the same time can also be necessary for arbitrage strategies, to benefit from the statistical discrepancies in the dynamics of two (or more) stocks – e.g., pair trading – or between the price of a basket of stocks and that of an index.

Knowing the optimal way to liquidate a multi-asset portfolio is also important for single-stock portfolios. A round trip on a second asset – typically a very liquid one, such as an index futures contract – can indeed be useful to partially hedge the price risk in the unwinding process of a single-stock portfolio. By doing so, one can slow down the execution process at the beginning and avoid paying too much in execution costs.

---

[3]This is not a real restriction when the market volume curve is flat, or, in practice, not too bumpy.

[4]The problem is indeed to minimize

$$\int_0^\tau \left(V_t L\left(\frac{q'(t)}{V_t}\right) + \frac{1}{2}\gamma\sigma^2 q(t)^2 \right) dt,$$

over the set

$$\mathcal{C}_{\rho_{\min},\tau} = \left\{ q \in W^{1,1}(0,\tau), q(0) = q_0, q(\tau) = 0, |q'(t)| \geq \rho_{\min} V_t \right\},$$

which is a convex set.

5.3.1 The model

We consider a portfolio with d different stocks. The initial position of the trader is denoted by $q_0 = (q_0^1, \ldots, q_0^d)$, where, for each asset i, q_0^i can be positive, negative, or equal to 0 (to cover the case of additional securities used only for hedging reasons). We consider the problem of unwinding this portfolio over the time interval $[0, T]$.

We denote by $(V_t^1)_{t \in [0,T]}, \ldots, (V_t^d)_{t \in [0,T]}$, the d market volume processes associated with the d stocks. As for the single-stock case, we assume that these market volume processes are deterministic, continuous, positive, and bounded.

The trader's position is modeled by a process $(q_t)_{t \in [0,T]} = (q_t^1, \ldots, q_t^d)_{t \in [0,T]}$ which has the following dynamics:

$$\forall i, dq_t^i = v_t^i dt, \tag{5.3}$$

where $(v_t)_{t \in [0,T]} = (v_t^1, \ldots, v_t^d)_{t \in [0,T]}$ is in the set $\mathcal{A}$ of admissible controls defined by

$$\mathcal{A} = \left\{ (v_t)_{t \in [0,T]} \in \mathbb{H}^0(\mathbb{R}^d, (\mathcal{F}_t)_t), \forall i, \int_0^T v_t^i dt = -q_0^i, \int_0^T |v_t^i| dt \in L^\infty(\Omega) \right\}.$$

For each stock, we consider that the dynamics of the mid-price is the same as in the single-stock case. In particular, the volume traded in one stock only affects the price of that stock:

$$\forall i, dS_t^i = \sigma^i dW_t^i + k^i v_t^i dt. \tag{5.4}$$

However, we do not assume independence between the Brownian motions $(W_t^1)_t, \ldots, (W_t^d)_t$. We suppose instead that the d-dimensional process $(\sigma^1 W_t^1, \ldots, \sigma^d W_t^d)_t$ has a nonsingular covariance matrix Σ.

Regarding the instantaneous component of market impact, we assume, as for the permanent component, that there is no cross-impact.[5] Therefore, we introduce d functions $L^1, \ldots, L^d$, which verify the same assumptions as in the single-asset case: strict convexity, nonnegativity, asymptotic super-linearity, and the assumption that there is no fixed cost.

The cash process $(X_t)_{t \in [0,T]}$ has therefore the following dynamics:

$$dX_t = \sum_{i=1}^d -v_t^i S_t^i dt - V_t^i L^i \left(\frac{v_t^i}{V_t^i} \right) dt. \tag{5.5}$$

---

[5]The case of cross-impacts is considered for instance in [162], but we believe that cross-impact is very hard to measure in practice. In any case it is expected to be significant in only a few situations.

In this section, we consider the problem of finding the maximizers of the objective function

$$\mathbb{E}\left[-\exp(-\gamma X_T)\right],$$

over the set $\mathcal{A}$ of admissible controls.

5.3.2   Towards a Hamiltonian system of $2d$ equations

To solve this problem, we first compute the expression of X_T using Eqs. (5.3), (5.4), and (5.5) and d integrations by parts.

$$
\begin{aligned}
X_T &= X_0 - \sum_{i=1}^{d} \int_0^T v_t^i S_t^i dt - \sum_{i=1}^{d} \int_0^T V_t^i L^i\left(\frac{v_t^i}{V_t^i}\right) dt \\
&= X_0 + \sum_{i=1}^{d} q_0^i S_0^i + \sum_{i=1}^{d} \int_0^T k^i v_t^i q_t^i dt \\
&\quad + \sum_{i=1}^{d} \int_0^T q_t^i \sigma^i dW_t^i - \sum_{i=1}^{d} \int_0^T V_t^i L^i\left(\frac{v_t^i}{V_t^i}\right) dt \\
&= X_0 + \sum_{i=1}^{d} q_0^i S_0^i - \sum_{i=1}^{d} \frac{k^i}{2} q_0^{i^2} \\
&\quad + \sum_{i=1}^{d} \int_0^T q_t^i \sigma^i dW_t^i - \sum_{i=1}^{d} \int_0^T V_t^i L^i\left(\frac{v_t^i}{V_t^i}\right) dt. \quad (5.6)
\end{aligned}
$$

We deduce from Eq. (5.6) that[6]

$$
\begin{aligned}
&\mathbb{E}\left[-\exp(-\gamma X_T)\right] \\
&= -\exp\left(-\gamma\left(X_0 + \sum_{i=1}^{d} q_0^i S_0^i - \sum_{i=1}^{d} \frac{k^i}{2} q_0^{i^2}\right)\right) \\
&\quad \times \mathbb{E}\left[\exp\left(\gamma \sum_{i=1}^{d} \int_0^T V_t^i L^i\left(\frac{v_t^i}{V_t^i}\right) dt - \gamma \sum_{i=1}^{d} \int_0^T q_t^i \sigma^i dW_t^i\right)\right] \\
&= -\exp\left(-\gamma\left(X_0 + \sum_{i=1}^{d} q_0^i S_0^i - \sum_{i=1}^{d} \frac{k^i}{2} q_0^{i^2}\right)\right) \\
&\quad \times \mathbb{E}^{\mathbb{Q}}\left[\exp\left(\gamma \sum_{i=1}^{d} \int_0^T V_t^i L^i\left(\frac{v_t^i}{V_t^i}\right) dt + \frac{1}{2}\gamma^2 \int_0^T q_t \cdot \Sigma q_t dt\right)\right],
\end{aligned}
$$

where $\cdot$ is the scalar product of $\mathbb{R}^d$, and where $\mathbb{Q}$ is the probability measure

---

[6]Hereafter, q is always a column vector.

defined by the following Radon-Nikodym derivative (the Girsanov theorem applies because q is bounded, by definition of $\mathcal{A}$):

$$\frac{d\mathbb{Q}}{d\mathbb{P}} = \exp\left(-\gamma \sum_{i=1}^{d} \int_0^T q_t^i \sigma^i dW_t^i - \frac{1}{2}\gamma^2 \int_0^T q_t \cdot \Sigma q_t dt\right).$$

By using the same reasoning as in Theorem 3.2, the problem boils down to finding a minimizer of

$$\int_0^T \left(\sum_{i=1}^{d} V_t^i L^i\left(\frac{v_t^i}{V_t^i}\right) + \frac{\gamma}{2} q_t \cdot \Sigma q_t\right) dt$$

over the set of deterministic controls $(v_t)_t \in \mathcal{A}$.

In other words, the problem boils down to a convex Bolza problem. We need therefore to find the minimizers of the functional J defined by

$$J(q) = \int_0^T \left(\sum_{i=1}^{d} V_t^i L^i\left(\frac{q^{i'}(t)}{V_t^i}\right) + \frac{\gamma}{2} q(t) \cdot \Sigma q(t)\right) dt,$$

over the set of $\mathbb{R}^d$-valued absolutely continuous functions $q \in W^{1,1}(0,T)$ satisfying the constraints $q(0) = q_0$ and $q(T) = 0$.

By using Theorems B.2 and B.3 of Appendix B, we can state the following result:

Theorem 5.3. *There exists a unique minimizer q^* of the function J over the set of $\mathbb{R}^d$-valued absolutely continuous functions $q \in W^{1,1}(0,T)$ satisfying the constraints $q(0) = q_0$ and $q(T) = 0$.*

Furthermore, q^* is uniquely characterized by

$$\begin{cases} p'(t) & = & \gamma \Sigma q^*(t), \\ q^{i*'}(t) & = & V_t^i H^{i'}(p^i(t)), \forall i, \\ q^*(0) & = & q_0, \\ q^*(T) & = & 0, \end{cases} \tag{5.7}$$

where $\forall i, H^i$ is the Legendre-Fenchel transform of the function L^i defined by

$$H^i(p) = \sup_\rho \rho p - L^i(\rho).$$

Theorem 5.3 gives a characterization of the optimal trading curves in the form of a Hamiltonian system with $2d$ equations – Eq. (5.7). This Hamiltonian system is made of a linear system of d equations ($p'(t) = \gamma \Sigma q^*(t)$), and d other equations, that are linear only in the case of quadratic execution cost functions. Eq. (5.7) generalizes Eq. (3.10), and we see that the interaction between assets occurs, as expected, through the variance-covariance matrix Σ. In particular, if asset prices are uncorrelated (i.e., if Σ is diagonal), then the Hamiltonian system of Eq. (5.7) is actually the set of the d Hamiltonian systems (similar to Eq. (3.10)) associated with the d unwinding processes of the d stocks, considered independently.

It is noteworthy that there is no monotonicity result, as far as trading curves are concerned, in the case of multi-asset portfolios. Even in the case of a long-only portfolio, it is possible to obtain an optimal liquidation strategy that consists in overselling before buying back. In other words, the trader may find it optimal to be short one or several stocks at some point in time over the course of the execution process, because it mitigates price risk (see Figure 6.4).

Another important point is that we can add participation constraints in the multi-asset case. If the set $\mathcal{A}$ of admissible controls is replaced by a set of the form

$$\mathcal{A}_{\rho_{\max}^1, \dots, \rho_{\max}^d} = \left\{ (v_t)_{t \in [0,T]} \in \mathbb{H}^0(\mathbb{R}, (\mathcal{F}_t)_t), \forall i, \int_0^T v_t^i dt = -q_0^i, |v_t^i| \le \rho_{\max}^i V_t^i \right\},$$

then under the assumption that $\forall i, |q_0^i| \le \rho_{\max}^i \int_0^T V_t^i dt$, there exists a unique optimal liquidation strategy q^* characterized by

$$\begin{cases} p'(t) &=& \gamma \Sigma q^*(t), \\ q^{i*'}(t) &=& V_t^i H_{\rho_{\max}^i}^{i'}(p^i(t)), \forall i, \\ q^*(0) &=& q_0, \\ q^*(T) &=& 0. \end{cases} \tag{5.8}$$

We shall see in Chapter 6 how to numerically approximate the solutions of all the Hamiltonian systems introduced so far. We will see, in particular, that the solutions of the Hamiltonian systems in the single-stock case are very easy to approximate numerically, but that the situation is more complex in the case of multi-asset portfolios.

5.3.3 How to hedge the risk of the execution process

The above multi-asset framework generalizes the Almgren-Chriss model to the case of complex portfolios. It can also be useful in the case of single-stock portfolios, when one wants to hedge the risk associated with the execution process. Instead of using the classical model of Chapter 3 to liquidate a position q_0 in a single asset, one can consider a second asset correlated to the first one, and then consider a multi-asset liquidation problem with an initial position equal to $(q_0, 0)$. Because of the correlation between the two assets, it is often optimal to do a round trip on the second asset in order to hedge the open position in the first one. This risk reduction has a cost, because the trader pays the execution costs related to the round trip on the second asset. However, it also enables to slow down the execution process for the first asset, and therefore to reduce the execution costs related to the first asset.

In practice, the risk of an execution process, be it for a single-asset portfolio or for a multi-asset portfolio, is often hedged with futures, and very rarely with another stock. In particular, one can assume that market impact and execution costs can be neglected, as far as the futures are concerned. In the single-stock case, if we consider the liquidation of a position q_0 in the stock, and if we add another asset with no execution costs – designated by the subscript h (for hedging) – the problem is no longer to minimize

$$J : q \mapsto \int_0^T \left(V_t L \left(\frac{q'(t)}{V_t} \right) + \frac{1}{2} \gamma \sigma^2 q(t)^2 \right) dt,$$

over

$$\mathcal{C} = \left\{ q \in W^{1,1}(0,T), q(0) = q_0, q(T) = 0 \right\},$$

but instead

$$J_h : (q, q_h) \mapsto$$

$$\int_0^T \left(V_t L \left(\frac{q'(t)}{V_t} \right) + \frac{1}{2} \gamma \left(\sigma^2 q(t)^2 + 2 \rho \sigma \sigma_h q(t) q_h(t) + \sigma_h^2 q_h(t)^2 \right) \right) dt,$$

over

$$\mathcal{C}_h = \left\{ (q, q_h) \in W^{1,1}(0,T) \times W^{1,1}(0,T), q(0) = q_0, q(T) = 0 \right\}.$$

It is straightforward to verify that the optimal hedge consists in setting $q_h(t) = -\rho \frac{\sigma}{\sigma_h} q(t)$. Therefore, the problem boils down to minimizing over $\mathcal{C}$ the new function

$$\begin{aligned} \tilde{J} : q \mapsto \tilde{J}(q) &= J_h \left(q, -\rho \frac{\sigma}{\sigma_h} q \right) \\ &= \int_0^T \left(V_t L \left(\frac{q'(t)}{V_t} \right) + \frac{1}{2} \gamma \sigma^2 (1 - \rho^2) q(t)^2 \right) dt. \end{aligned}$$

In other words, when the execution costs on the second asset can be neglected, the optimal trading curve $t \mapsto q^*(t)$ for the first asset is the same as if there was no hedge, but a value of the volatility parameter equal to $\sigma \sqrt{1 - \rho^2}$ instead of σ. Regarding the second asset, the optimal strategy consists in building an initial position $-\rho \frac{\sigma}{\sigma_h} q_0$ (this is done instantaneously, since we assume that there is no friction on this asset), and then liquidating it progressively with an optimal trading curve given by $q_h^*(t) = -\rho \frac{\sigma}{\sigma_h} q^*(t)$.

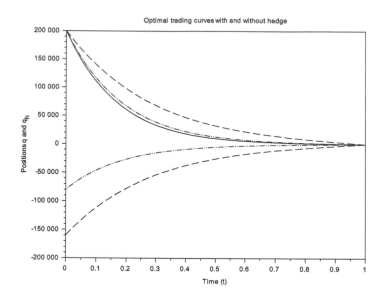

FIGURE 5.1: Optimal trading curve for $q_0 = 200,000$ shares over one day $(T = 1)$, for $\gamma = 5.10^{-6}$ €$^{-1}$, with and without hedge. Solid line: without hedge. Dash-dotted lines: optimal strategy when $\sigma_h = \sigma$ and $\rho = 0.4$. Dashed lines: optimal strategy when $\sigma_h = \sigma$ and $\rho = 0.8$. Decreasing curves correspond to q, whereas increasing ones correspond to q_h.

Figure 5.1 represents the optimal trading curve with and without hedge, for a stock that has the following characteristics:

- $S_0 = 45$ €,

- $\sigma = 0.6$ €·day$^{-1/2}$·share^{-1}, i.e., an annual volatility approximately equal to 21%,

- $V = 4,000,000$ shares·day^{-1},

- $L(\rho) = \eta\rho^2$, with $\eta = 0.1$ €·share^{-1}.

We see that the higher the correlation between the two asset prices, the closer to the straight line the optimal trading curve for the first asset. This is natural, because the round trip strategy on the asset used to hedge removes part of the risk associated with the execution process, and it makes it possible therefore to slow down the execution process for the first asset.

5.4 Conclusion

In this chapter we have presented several extensions of the Almgren-Chriss model. Considering a nontrivial dynamics for the stock price is interesting to account for a drift, or to take account of the seasonality effects of intraday volatility. We have seen that there was no additional difficulty associated with these extensions. Constraining the participation rate to be below a certain threshold is often required in practice, and we have seen that there is no difficulty in finding the Hamiltonian system associated with the constrained optimization problem. The multi-asset case is also important in many situations, from Program Trading to the hedging of an execution process. We have seen how to derive the Hamiltonian system in the general multi-asset case. We have also dealt with the limit case – very important in practice – in which the execution costs on the hedging asset can be neglected.

These generalizations have been presented independently of one another in the above text. However, it is straightforward to notice that the extensions presented in this chapter can be included altogether into a single model.

So far, in numerical examples, we have only considered the case of quadratic execution cost functions. To go beyond this quadratic case, we need numerical methods for approximating the solution of the Hamiltonian systems presented in Chapters 3, 4, and 5. The next chapter is dedicated to numerical methods. We will see that, in the case of single-stock portfolios, very simple shooting methods make it possible to approximate rapidly the solution of the Hamiltonian systems (3.10), (4.17), (5.1), and (5.2), for all kinds of execution functions L. However, when it comes to the liquidation of multi-asset portfolios, the most relevant numerical method depends on the regularity of the execution functions L^is, and of their convex conjugates H^is, in Eq. (5.7) or Eq. (5.8).

Chapter 6

Numerical methods

Computers are useless. They can only give you answers.

— Pablo Picasso

A man provided with paper, pencil, and rubber, and subject to strict discipline, is in effect a universal machine.

— Alan Turing

In the previous chapters, we saw that, at least when using the Almgren-Chriss framework, the optimal trading curve for almost all execution problems was characterized by a Hamiltonian system. In this chapter, we present several numerical methods to approximate the solution of these Hamiltonian systems. For single-asset portfolios, the method we propose is a very simple and efficient shooting method. For more complex portfolios, we propose several methods, depending on the type of execution cost functions.

6.1 The case of single-stock portfolios

We have seen, in Chapters 3 and 4, that for all common types of orders (IS, VWAP, TC, etc.[1]), the optimal trading curve, as computed in the Almgren-Chriss model, was characterized by a Hamiltonian system. However, except in some specific cases (mainly in the case of a quadratic execution cost function and a flat market volume curve – see Eq. (3.14) for instance), the optimal trading curve cannot be computed in closed-form. We therefore need to rely on numerical methods for approximating the optimal trading curve.

In this section, we consider the case of an IS order to liquidate a single-stock portfolio. Moreover, to take account of one of the extensions presented in

---

[1]Not POV orders, because they are too simple.

Chapter 5, we assume that there is an upper bound $\rho_{\max}$ for the participation rate to the market. In other words, we focus on the Hamiltonian system (5.2) recalled below:

$$\begin{cases} p'(t) &=& \gamma\sigma^2 q^*(t), \\ q^{*'}(t) &=& V_t H'_{\rho_{\max}}(p(t)), \\ q^*(0) &=& q_0, \\ q^*(T) &=& 0, \end{cases}$$

where

$$H_{\rho_{\max}}(p) = \sup_{|\rho| \le \rho_{\max}} \rho p - L(\rho)$$

is a C^1 function, because L was assumed to be strictly convex in Chapter 3.

Our goal in this section is to present a numerical method for approximating q^*, the solution of the above system.[2] The method we propose could be used in the same way for the other Hamiltonian systems found in Chapters 3, 4, and 5 – at least in the case of single-stock portfolios.

6.1.1   A shooting method

The method we propose is a shooting method. The first step consists in discretizing the above system. For this purpose, we use the discretization corresponding to the discrete-time counterpart of the Almgren-Chriss model presented in Section 3.3.

We consider a subdivision $t_0 = 0 < \ldots < t_n = n\Delta t < \ldots < t_N = N\Delta t = T$ of the time interval $[0, T]$ into N slices of length Δt, and we want to find a solution $(q_0^*, \ldots, q_N^*, p_0, \ldots, p_{N-1})$ to the system

$$\begin{cases} p_{n+1} &=& p_n + \Delta t \gamma \sigma^2 q_{n+1}^*, & 0 \le n < N-1, \\ q_{n+1}^* &=& q_n^* + \Delta t V_{n+1} H'_{\rho_{\max}}(p_n), & 0 \le n < N, \end{cases} \qquad q_0^* = q_0, \quad q_N^* = 0.$$

$$(6.1)$$

The second step of the method consists in replacing the system (6.1) by another system, where the constraint on the terminal value q_N is relaxed, and replaced by an initial value $p_0 = \lambda$. In other words, for $\lambda \in \mathbb{R}$, we consider the system

$$\begin{cases} p_{n+1}^\lambda &=& p_n^\lambda + \Delta t \gamma \sigma^2 q_{n+1}^\lambda, & 0 \le n < N-1, \\ q_{n+1}^\lambda &=& q_n^\lambda + \Delta t V_{n+1} H'_{\rho_{\max}}(p_n^\lambda), & 0 \le n < N, \end{cases} \qquad q_0^\lambda = q_0, \quad p_0^\lambda = \lambda.$$

$$(6.2)$$

The recursive system (6.2) can be solved very easily, because it is a purely forward one.

---

[2]We do not really care about p. It is noteworthy that, although q^* is unique, p is not – see Section 3.2.3.2.

Then, the shooting method aims at finding a value of λ such that the solution (p^λ, q^λ) of Eq. (6.2) verifies $q_N^\lambda = 0$, and is therefore a solution of Eq. (6.1). Because Eq. (6.1) has a solution, we know that there exists such a λ. The challenge is to find it numerically.

For this purpose, we rely on the following result:

Proposition 6.1. *Let (p^λ, q^λ) be a solution of Eq. (6.2). Then,*

$$\lambda \mapsto q_N^\lambda$$

is a continuous and nondecreasing function.

Proof. *By definition, $\lambda \mapsto p_0^\lambda = \lambda$ and $\lambda \mapsto q_0^\lambda = q_0$ are two continuous and nondecreasing functions.*

Let us consider $n \in \{0, \ldots, N-2\}$, and assume that $\lambda \mapsto p_n^\lambda$ and $\lambda \mapsto q_n^\lambda$ are two continuous and nondecreasing functions.

Because $H_{\rho_{max}}$ is convex, $H'_{\rho_{max}}$ is a nondecreasing function (it is also continuous by assumption), and therefore

$$\lambda \mapsto q_{n+1}^\lambda = q_n^\lambda + \Delta t V_{n+1} H'_{\rho_{max}}(p_n^\lambda)$$

is a continuous and nondecreasing function. Now, by definition,

$$\lambda \mapsto p_{n+1}^\lambda = p_n^\lambda + \Delta t \gamma \sigma^2 q_{n+1}^\lambda$$

is also continuous and nondecreasing.

We have proved therefore, by induction, that $\forall n \in \{0, \ldots, N-1\}$, $\lambda \mapsto p_n^\lambda$ and $\lambda \mapsto q_n^\lambda$ are two continuous and nondecreasing functions.

In particular, for $n = N-1$, $\lambda \mapsto p_{N-1}^\lambda$ and $\lambda \mapsto q_{N-1}^\lambda$ are continuous and nondecreasing, and therefore

$$\lambda \mapsto q_N^\lambda = q_{N-1}^\lambda + \Delta t V_{n+1} H'_{\rho_{max}}(p_{N-1}^\lambda)$$

is a continuous and nondecreasing function.

Using Proposition 6.1, we see that the shooting method boils down to finding a root to the continuous and nondecreasing function $\lambda \mapsto q_N^\lambda$. Therefore, we can use the bisection method to find a value λ^* such that $q_N^{\lambda^*} \simeq 0$. Then, the solution $(q_0^*, \ldots, q_N^*)$ of Eq. (6.1) is well approximated by $(q_0^{\lambda^*}, \ldots, q_N^{\lambda^*})$.

6.1.2   Examples

In Chapter 3, we computed several examples of trading curves in a very specific case: quadratic execution costs, a flat market volume curve, and no constraint on the participation rate.

Using the above numerical method, we can consider some more realistic cases, where $\phi \in (0,1)$.[3] Furthermore, we can take account of intraday seasonality in market volume curves, and we can add a constraint on the participation rate very easily.

In this section, we consider the case of a stock with the following characteristics:

- $S_0 = 45$ €,

- $\sigma = 0.6$ €·day$^{-1/2}$·share^{-1}, i.e., an annual volatility approximately equal to 21%,

- The market volume curve corresponds to a total market daily volume of 4,000,000 shares, with the relative market volume curve of Figure 4.3. This market volume curve is denoted by $(V_t)_t$.

- $L(\rho) = \eta|\rho|^{1+\phi} + \psi|\rho|$, where $\eta = 0.1$ €·share^{-1}, $\psi = 0.004$ €·share^{-1}, and $\phi = 0.75$.[4]

We consider the case of a trader who has to liquidate $q_0 = 200{,}000$ shares over the continuous auction period of one given day ($T = 1$). This number of shares represents 5% of the total daily volume.

We see on Figure 6.1 that the shape of the optimal trading curve is influenced by the shape of the market volume curve. We see in particular that the trader executes faster at the beginning and at the end of the day, but also around 11:30, in line with the relative market volume curve of Figure 4.3.

In Figure 6.2, we have plotted the optimal trading curves for three different values of γ, when $\rho_{\max}$ is set to a value such that the constraint is never binding. We see, as expected, that the more risk-averse the trader, the faster he liquidates the position. Like in Chapter 3, we see that the choice of γ is crucial. In particular, we notice that the choice of γ has more influence on the optimal trading curve than the exact shape of the market volume curve does.[5]

---

[3]We refer to Chapter 7, or to [10], for a method to estimate execution costs and market impact parameters.

[4]We recall that ψ plays no role in the single-asset case.

[5]Methods to choose γ are discussed in Sections 4.3.3 and 8.3.3.

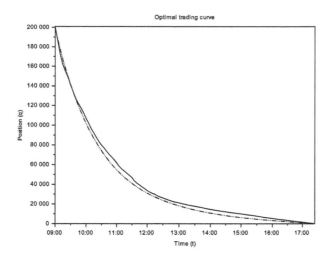

FIGURE 6.1: Optimal trading curve for $q_0 = 200,000$ shares over one day $(T = 1)$, for different market volume curves. Solid line: market volume curve $(V_t)_t$. Dash-dotted line: flat market volume curve with 4,000,000 shares per day – $\gamma = 5.10^{-6}$ €$^{-1}$, $\rho_{\max} = 5$, so that the constraint is never binding.

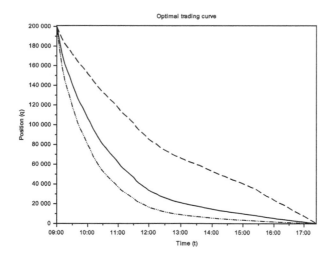

FIGURE 6.2: Optimal trading curve for $q_0 = 200,000$ shares over one day $(T = 1)$, for different values of γ. Dash-dotted line: $\gamma = 10^{-5}$ €$^{-1}$. Solid line: $\gamma = 5.10^{-6}$ €$^{-1}$. Dashed line: $\gamma = 10^{-6}$ €$^{-1}$ – $\rho_{\max} = 5$, as above.

The predominance of the risk aversion parameter γ over the other parameters is an important feature of the Almgren-Chriss model, and it may be seen as an issue because the choice of γ is always approximative. In the case of quadratic execution costs, we have seen in Eq. (3.14) that the influence of the variables happens through the ratio $\frac{\gamma \sigma^2 V}{2\eta}$. More generally, for an execution cost function $L(\rho) = \eta |\rho|^{1+\phi}$, we see from Eqs. (3.11) and (3.12) that the relevant variable is in fact $\frac{\gamma \sigma^2 V}{(1+\phi)^{\frac{1}{\phi}} \eta^{\frac{1}{\phi}}}$.

Most of the parameters can be estimated: the volatility parameter σ, the liquidity parameters η and ϕ, and the market volume V (in fact the market volume curve $(V_t)_t$). The estimations are always subject to errors, but, for a given stock, the value of a parameter can be estimated with confidence inside a small range. As far as γ is concerned, the situation is different: the choice is usually subject to a factor 2, or up to 5. This is why γ plays such an important role in the determination of optimal trading curves.

In Figure 6.3, we have plotted the optimal trading curves for four different values of $\rho_{\max}$, when $\gamma = 5.10^{-6} \ \text{€}^{-1}$. When $\rho_{\max}$ is equal to 20%, 15%, or 10%, we see that the constraint is binding at the beginning of the liquidation process, that is, when the trader would like to execute faster.

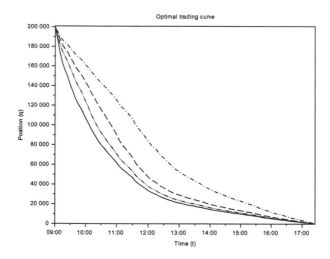

FIGURE 6.3: Optimal trading curve for $q_0 = 200{,}000$ shares over one day ($T = 1$), for different values of $\rho_{\max}$. Solid line: $\rho_{\max} = 5$ (a very high value, such that the constraint is never binding). Dash-dotted line (two dots): $\rho_{\max} = 20\%$. Dashed line: $\rho_{\max} = 15\%$. Dash-dotted line (one dot): $\rho_{\max} = 10\%$.

The influence of the participation constraint $\rho_{\max}$ turns out to be very important. Even for an order that represents 5% of the volume traded over the period, it makes a real difference to choose $\rho_{\max} = 20\%$, $\rho_{\max} = 15\%$, or $\rho_{\max} = 10\%$.

6.1.3 Final remarks on the single-asset case

The above shooting method is very efficient to approximate numerically the optimal trading curve in the case of IS orders, Target Close orders, or VWAP orders (with permanent market impact). It can be used with any shape of market volume curve, and even when a (maximum) constraint on the participation rate to the market is imposed. Furthermore, it can be used even when one adds a drift in the price process or considers a nonconstant (but deterministic) volatility process, as in Chapter 5. In all cases, it is a very fast method.

However, the shooting method we have presented is based on a monotonicity result (Proposition 6.1), and this result is specific to dimension 1. In short, the above shooting method cannot be generalized to tackle execution problems in the case of multi-asset portfolios.

6.2 The case of multi-asset portfolios

We now consider the case of a portfolio with d stocks, which must be liquidated over a time interval $[0, T]$. We saw in Chapter 5 that the Almgren-Chriss framework could be extended to multi-asset portfolios. With the same notations as in Chapter 5, we know that the d optimal trading curves are characterized by a Hamiltonian system of the form

$$\begin{cases} p'(t) &= \gamma \Sigma q^*(t), \\ q^{i*'}(t) &= V_t^i H^{i'}(p^i(t)), \quad 1 \le i \le d, \\ q^*(0) &= q_0, \\ q^*(T) &= 0, \end{cases}$$

where $H^i(p) = \sup_\rho \rho p - L^i(\rho)$.

This Hamiltonian system can also be generalized to include maximum constraints on the participation rate for each stock – see Eq. (5.8).

As for single-asset portfolios, the first step consists in discretizing the system. We consider a subdivision $t_0 = 0 < \ldots < t_n = n\Delta t < \ldots < t_N = T$ of the time interval $[0, T]$.

We want to find a solution $\left((q_n^{i*})_{1\leq i\leq d,0\leq n\leq N},(p_n^i)_{1\leq i\leq d,0\leq n\leq N-1}\right)$ to the following system[6]

$$
\begin{cases}
p_{n+1} = p_n + \Delta t\gamma\Sigma q_{n+1}^*, & 0\leq n<N-1,\\
q_{n+1}^{i*} = q_n^{i*} + \Delta t V_{n+1}^i H^{i'}(p_n^i), & 0\leq n\leq N-1, 1\leq i\leq d,\\
q_0^* = q_0 \in \mathbb{R}^d,\\
q_N^* = 0 \in \mathbb{R}^d.
\end{cases}
\tag{6.3}
$$

6.2.1 Newton's method for smooth Hamiltonian functions

The first method we propose for solving the problem in the multi-asset case is a Newton's method.

We can indeed regard Eq. (6.3) as an equation of the form

$$
F\left((q_n^{i*})_{1\leq i\leq d,1\leq n\leq N-1},(p_n^i)_{1\leq i\leq d,0\leq n\leq N-1}\right)=0,
$$

and apply the classical Newton's method to the function F. However, for the classical Newton's method to apply, we need to have $H^{i'}$ differentiable[7] for all i. This is the case, for instance, when $L^i(\rho)=\eta^i|\rho|^{1+\phi^i}$, $\phi^i\in(0,1]$.

In the specific case of the Hamiltonian system (6.3), the Newton's method consists in starting with a first guess

$$
\left((q_n^{0,i})_{1\leq i\leq d,0\leq n\leq N},(p_n^{0,i})_{1\leq i\leq d,0\leq n\leq N-1}\right),
$$

hopefully not too far from the actual solution, and then to define iteratively

$$
\left((q_n^{k,i})_{1\leq i\leq d,0\leq n\leq N},(p_n^{k,i})_{1\leq i\leq d,0\leq n\leq N-1}\right),
$$

for $k\geq 0$ by

- $q_0^{k+1,i}=q_0^i$, $q_N^{k+1,i}=0$, $1\leq i\leq d$,

- for $1\leq i\leq d$ and $1\leq n\leq N-1$,

$$
q_n^{k+1,i}=q_n^{k,i}+\delta q_n^{k+1,i},
$$

- for $1\leq i\leq d$ and $0\leq n\leq N-1$,

$$
p_n^{k+1,i}=p_n^{k,i}+\delta p_n^{k+1,i},
$$

where for $0\leq n<N-1$,

$$
\delta p_{n+1}^{k+1,\cdot}=\delta p_n^{k+1,\cdot}+\Delta t\gamma\Sigma\delta q_{n+1}^{k+1,\cdot}-\left(p_{n+1}^{k,\cdot}-p_n^{k,\cdot}-\Delta t\gamma\Sigma q_{n+1}^{k,\cdot}\right)
\tag{6.4}
$$

---

[6]p_n and q_n designate column vectors with d coordinates corresponding to the d different assets.

[7]There exist other methods that generalize the classical Newton's method by using the generalized Jacobian of Clarke. We do not consider these methods here.

and, for $0 \leq n < N - 1, 1 \leq i \leq d$,

$$\delta q_{n+1}^{k+1,i} = \delta q_n^{k+1,i} + \delta p_n^{k+1,i} \Delta t V_{n+1}^i H^{i''}(p_n^{k,i})$$

$$- \left(q_{n+1}^{k,i} - q_n^{k,i} - \Delta t V_{n+1}^i H^{i'}(p_n^{k,i}) \right). \tag{6.5}$$

Eqs. (6.4) and (6.5) constitute a linear system of equations, which can be solved very easily. Therefore, one can compute iteratively the values of the sequence $\left((q_n^{k,i})_{1 \leq i \leq d, 0 \leq n \leq N}, (p_n^{k,i})_{1 \leq i \leq d, 0 \leq n \leq N-1} \right)_k$, and hope for the Newton's method to converge.

As far as the initial guess is concerned, a smart choice consists in considering a vector

$$\left((q_n^{0,i})_{1 \leq i \leq d, 1 \leq n \leq N-1}, (p_n^{0,i})_{1 \leq i \leq d, 0 \leq n \leq N-1} \right)$$

satisfying

$$q_0^{0,i} = q_0^i, q_N^{0,i} = 0, \quad 1 \leq i \leq d,$$

and

$$p_{n+1}^{0,\cdot} = p_n^{0,\cdot} + \Delta t \gamma \Sigma q_{n+1}^{0,\cdot}, \quad 0 \leq n < N - 1.$$

By immediate induction, we have indeed that, for all $k \geq 0$,

$$p_{n+1}^{k,\cdot} = p_n^{k,\cdot} + \Delta t \gamma \Sigma q_{n+1}^{k,\cdot}, \quad 0 \leq n < N - 1,$$

and this simplifies the above system. In practice, it is for instance possible to consider as a first guess the solution obtained when one replaces all the parameters $\phi^1, \ldots, \phi^d$ by 1, because in that case the system (6.3) is linear and can be solved numerically very easily.

The above Newton's method works quite well in many cases, but, as for most Newton's methods, it requires to start with an initial guess that is not far from the actual solution. Furthermore, the Newton's method only applies when the Hamiltonian functions are sufficiently smooth; this is a problem when considering a bid-ask spread component or a constraint on the participation rate.

6.2.2 Convex duality to the rescue

Many methods coming from convex optimization can be used to solve numerically the problem. A general one has been developed in Guéant, Lasry, and Pu [85]. We refer the reader to the paper for details on the method, but we want to state that the dual problem can be very helpful when it comes to solving the problem numerically. The dual problem[8] associated with the system (6.3) is the minimization of the function J_d defined by[9]

$$J_d((p_n^i)_{0 \leq n < N, 1 \leq i \leq d}) =$$

$$\sum_{i=1}^{d} \sum_{n=0}^{N-1} V_{n+1}^i H^i(p_n^i) \Delta t + \frac{1}{2\gamma\Delta t} \sum_{n=1}^{N-1} (p_n - p_{n-1}) \Sigma^{-1} (p_n - p_{n-1}) + \sum_{i=1}^{d} p_0^i q_0^i.$$

Because the functions L^is are strictly convex, the Hamiltonian functions H^is (or the functions $\left(H^i_{\rho^i_{\max}} \right)_i$, if there is a participation constraint) are C^1, and a gradient descent is always possible to approximate a minimizer $(p_n^i)_{0 \leq n < N, 1 \leq i \leq d}$ of J_d, and then approximate the optimal trading curves using the relation

$$q_{n+1}^* = \frac{1}{\gamma\Delta t} \Sigma^{-1}(p_{n+1} - p_n), \quad 0 \leq n < N - 1$$

between the primal and dual variables.

In [85], the authors propose a modified gradient descent method based on the dual optimization problem, and show that their method converges whenever the first derivative of each Hamiltonian function is Lipschitz – this condition is verified by all the functional forms used in practice.

6.2.3   Examples

We now present a few examples in the case of multi-asset portfolios. For the sake of simplicity, we consider in particular portfolios with two stocks.

The first stock (Asset 1) has the following characteristics:

- $S_0 = 100$ €,

- $\sigma = 1.2$ €·day$^{-1/2}$·share^{-1}, i.e., an annual volatility approximately equal to 19%,

- The market volume curve is flat with $V = 3,000,000$ shares·day^{-1},

- $L(\rho) = \eta|\rho|^{1+\phi} + \psi|\rho|$, where $\eta = 0.5$ €·share^{-1}, $\phi = 0.5$, and $\psi = 0.01$ €·share^{-1}.

---

[8]See Theorem B.4 of Appendix B.
[9]When a maximum participation rate is imposed, the function J_d is instead defined by

$$J_d((p_n^i)_{0 \leq n < N, 1 \leq i \leq d}) =$$

$$\sum_{i=1}^{d} \sum_{n=0}^{N-1} V_{n+1}^i H^i_{\rho^i_{\max}}(p_n^i) \Delta t + \frac{1}{2\gamma\Delta t} \sum_{n=1}^{N-1} (p_n - p_{n-1}) \Sigma^{-1} (p_n - p_{n-1}) + \sum_{i=1}^{d} p_0^i q_0^i.$$

The second stock (Asset 2) has the following characteristics:

- $S_0 = 45$ €,

- $\sigma = 0.6$ €·day$^{-1/2}$·share^{-1}, i.e., an annual volatility approximately equal to 21%,

- The market volume curve is flat with $V = 4{,}000{,}000$ shares·day^{-1},

- $L(\rho) = \eta|\rho|^{1+\phi} + \psi|\rho|$, where $\eta = 0.1$ €·share^{-1}, $\phi = 0.75$, and $\psi = 0.004$ €·share^{-1}.

In Figure 6.4, we consider the liquidation of a portfolio with $q_0^1 = 200{,}000$ shares of the first stock and $q_0^2 = 100{,}000$ shares of the second stock (with $T = 1$). We assume that the correlation between the price increments of the two stocks is 80%. Moreover, the maximum participation rate for the first stock is $\rho_{\max}^1 = 25\%$, while there is no maximum participation rate for the second stock. In this example, the risk aversion parameter is $\gamma = 5.10^{-6}$ €$^{-1}$.

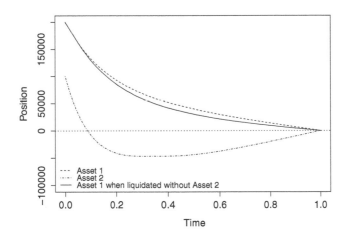

FIGURE 6.4: Optimal trading curves for a two-stock portfolio (1).

We see in Figure 6.4 that the 100,000 shares of the second stock are liqui-dated very rapidly. Then, a short position of ∼ 50,000 shares is built, and it is finally progressively unwound. In fact, the initial position in the second stock is smaller than the initial position in the first stock. Furthermore, the second stock is more liquid than the first one. Therefore, by using the same reasoning as in Section 5.3.3, the second stock is used to hedge the risk associated with the liquidation process of the first stock. It may seem counterintuitive, when one liquidates a long-only portfolio, to oversell one asset before buying it back. However, this is natural because of the risk associated with the liquidation of the important position in the (illiquid) Asset 1.

In Figure 6.5, we consider the case of a portfolio with $q_0^1 = 200{,}000$ shares of the first stock and $q_0^2 = 300{,}000$ shares of the second stock (with $T = 1$). We assume that the correlation between the price increments of the two stocks is negative, and equal to -20%. As above, the risk aversion parameter is $\gamma = 5.10^{-6}\ \text{€}^{-1}$, and we assume that there is a maximum participation rate of $\rho_{\max}^1 = 25\%$ for the first stock, and no maximum participation rate for the second stock.

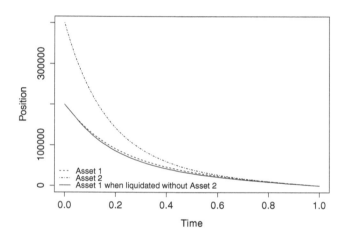

FIGURE 6.5: Optimal trading curves for a two-stock portfolio (2).

In this second example, the two assets are negatively correlated and the hedge is therefore automatic. In particular, we see that the presence of Asset 2 makes it possible to slow down the liquidation process of Asset 1.

The above numerical examples highlight that optimal execution strategies in the case of multi-asset portfolios are very sensible to the value of correlation coefficients. This is an important limitation of our model in terms of robustness.

Estimating daily correlations of prices with time series of market prices is a straightforward exercise. However, forecasting with confidence the correlation of intraday price increments is a difficult task. When one uses our model (or, in fact, any model with intraday correlations), the question of the robustness with regard to changes in the correlation parameters should be raised. In particular, it is important to estimate the cost of using a strategy expected to be optimal, when the real values of the correlation parameters eventually depart from the ones they were assigned to design the trading curves.[10]

6.3 Conclusion

The literature on optimal execution seldom tackles numerical methods.[11] In fact, the equations can easily be solved numerically in the case of the liquidation of a single-stock portfolio. However, the problem is not trivial for multi-asset portfolios. In the general case of the liquidation of a multi-asset portfolio, with general execution cost functions and maximum participation rates, it is often convenient to solve numerically the dual problem, by using the various techniques of convex optimization. For instance, a modified gradient descent method based on the dual problem is presented in [85].

Once optimal trading curves have been computed – numerically most of the time – it is of the utmost importance to go beyond the optimal scheduling problem, and to decide how to proceed to follow, in practice, the optimal trading curve. The next chapter mainly addresses the question of the relevant types of models for designing optimal trading tactics.

---

[10]Robust control techniques can be used for tackling this issue. In the two-asset case, instead of minimizing a function of the form

$$\int_0^T \left(V_t^1 L^1 \left(\frac{q^{1'}(t)}{V_t^1} \right) + V_t^2 L \left(\frac{q^{2'}(t)}{V_t^2} \right) \right.$$

$$\left. + \frac{1}{2}\gamma \left((\sigma^1)^2 q^1(t)^2 + 2\rho\sigma^1\sigma^2 q^1(t)q^2(t) + (\sigma^2)^2 q^2(t)^2 \right) \right) dt,$$

one could minimize a function of the form

$$\sup_{\rho \in [\underline{\rho}, \overline{\rho}]} \int_0^T \left(V_t^1 L^1 \left(\frac{q^{1'}(t)}{V_t^1} \right) + V_t^2 L \left(\frac{q^{2'}(t)}{V_t^2} \right) \right.$$

$$\left. + \frac{1}{2}\gamma \left((\sigma^1)^2 q^1(t)^2 + 2\rho\sigma^1\sigma^2 q^1(t)q^2(t) + (\sigma^2)^2 q^2(t)^2 \right) \right) dt,$$

where ρ is the correlation coefficient between the two assets, and where $[\underline{\rho}, \overline{\rho}]$ is an interval in which ρ is expected to lie.

[11]The paper [122] of Labadie and Lehalle is an exception.

Chapter 7

Beyond Almgren-Chriss

Plus c'est savant, plus c'est bête.

— Witold Gombrowicz

In this chapter, we analyze the evolution of the academic literature on optimal execution beyond the framework proposed by Almgren and Chriss. We discuss more general market impact models, the use of limit orders and dark pools, etc. We also focus on the modern understanding of the Almgren-Chriss model as a model for optimally scheduling the execution process, not as a model to execute Implementation Shortfall orders with market orders only. In particular, we discuss the two-layer approach of most execution algorithms, and the way the recent literature on optimal execution can help to design execution algorithms/tactics aimed at following the optimal trading curves computed in models *à la* Almgren-Chriss. Furthermore, we discuss market impact estimation. In particular, we highlight the fact that market impact and execution cost estimates used in execution algorithms should be based on execution data rather than market data.

7.1   Overview of the literature

The original model of Almgren and Chriss [8, 9] aimed at deriving the optimal way to execute a large (IS) order with market orders. We have already highlighted in Chapter 3 that the Almgren-Chriss model is nowadays used as a model to optimally schedule an execution process (as a whole), not as an execution model involving market orders only. However, the evolution of the academic literature has been linked to the seminal papers [8, 9], and to their initial understanding by the academic community.

This literature can be divided into three parts focusing on three different sets of limitations or assumptions of the original Almgren-Chriss model.

7.1.1 Models with market orders and the Almgren-Chriss market impact model

The first part of the literature is made of basic extensions of the initial Almgren-Chriss framework.

Almgren proposed for instance in [6] a model in continuous time, where execution costs have a random component. He also proposed a model to take stochastic volatility and liquidity into account – see [7]. Instead of the initial Bachelier (normal) dynamics for the stock price, a Black-Scholes (log-normal) dynamics has also been considered in the literature (see for instance Gatheral and Schied [75]). Numerous extensions of this kind have been proposed. They are interesting from a theoretical point of view, but except for a few of them (e.g., [6]), the extensions proposed in this first part of the literature lead to execution strategies that are not in the form of a trading curve.

Considering stochastic liquidity, stochastic volatility, or imposing a non-Brownian dynamics for the stock price leads indeed to optimal trading strategies that are genuine dynamic strategies. This is not a problem, but is there a real interest in considering stochastic volatility or a more realistic dynamics for the price process? Do these extensions provide a gain that offsets the cost incurred by the increased complexity in going from a trading curve computed *ex-ante* to a dynamic strategy? These questions are obviously not addressed in the literature, but they should be answered by anybody willing to use a model of this type in practice.

In this first part of the literature, there are also many articles dealing with the different objective functions that could be used, instead of the expected CARA utility function introduced in Chapter 3.[1] For instance, in [71] and [169], the authors proposed to use a mean-quadratic-variation criterion instead of a mean-variance one. More interestingly, the general case of IARA and DARA utility functions was considered by Schied and Schöneborn in [161], in a model where the time horizon to liquidate the portfolio is infinite. Schied and Schöneborn showed for instance in [161] that IARA utility functions are associated with aggressive in-the-money strategies, whereas DARA utility functions are associated with passive in-the-money strategies.[2]

The problem with these interesting extensions is twofold. First, they often lead to optimal execution strategies that are dynamic. Therefore, they raise the same questions as above. Second, for all relevant criterions, one needs to

---

[1]We recall that this objective criterion boils down to a mean-variance one, and that Almgren and Chriss initially considered a mean-variance criterion.

[2]This means in particular that CARA utility functions lead to strategies that are at the same time aggressive in-the-money and passive in-the-money, and therefore independent of the stock price, as in Chapter 3.

choose the value of one or several parameters linked to risk aversion. In the case of CARA utility functions, we have seen in Chapter 4 (see also Chapter 8) how the absolute risk aversion parameter γ could be chosen. For other utility functions or other criterions, the literature on optimal execution does not say much about how to choose the parameter(s). Furthermore, we have seen in Chapter 3 that, when a CARA utility function is used, the value of γ could be renormalized to depart from a strict CARA utility function setting. Therefore, in practice, there is little incentive in not using CARA utility functions and mean-variance criterions.

This first part of the literature exists for obvious reasons: when a new model emerges, researchers (and I am one of them) try to generalize it by adding some new features. However, as the great mathematician George Pólya said: "a mathematician who can only generalise is like a monkey who can only climb up a tree." Climbing down the tree is also necessary, and there was even a need to go back to the roots for questioning the two main assumptions of the Almgren-Chriss model (in its initial understanding): the market impact model, which is very/too simple, and the use of market orders only.

7.1.2 Models with transient market impact

The second part of the literature is related to market impact modeling. The Almgren-Chriss model, seen as a model to execute a large order with market orders, has often been criticized for the lack of realism of its market impact model. Subsequently, researchers addressed the question of the optimal execution strategies when the market impact is assumed to be transient.

The first paper in that direction is the paper [151] of Obizhaeva and Wang.[3] As in the original models of Almgren and Chriss, they considered execution with market orders only. However, they assumed that, when a market order hits the limit order book, the latter – and therefore the price – is immediately impacted by the trade, and it progressively returns to its initial state, or to a slightly different state when permanent market impact is taken into account. Their model features a flat limit order book and an exponentially decaying market impact. It was then generalized, and models with transient market impact (or equivalently resilient limit order books), are now numerous – see for instance [3], [4], [76], or the recent paper [56].

Gatheral introduced a general formulation for transient market impact models in [74]. With the notations of Chapter 3, the stock price dynamics

---

[3] Obizhaeva and Wang's article [151] was published in 2013, but it has been a working paper on the Internet since 2005.

in [74] is

$$S_t = S_0 + \int_0^t f(v_s)G(t-s)ds + \sigma W_t,$$

where the function f scales the immediate impact of trades as a function of the volume traded, and where the function G (called the decay kernel) models the decay of this immediate impact.[4] This formulation encompasses permanent market impact (when G is a constant function, or when $\lim_{+\infty} G > 0$), instantaneous market impact (when G is the Dirac mass δ_0), and whatever is in between, that is, transient market impact.

The main problem with these models is that many forms of transient market impact lead to the existence of price manipulation strategies. In the case of a limit order book assumed to be flat, i.e., f linear, it was proved in [5] that the market impact should decay as a convex nonincreasing function of time in order to prevent such price manipulation strategies. However, the situation is more complex in models where limit order books are not assumed to be flat – see [74] for necessary conditions in the case of a power function f and a power-law decay kernel G.

Although the models of this second part of the literature are all models of optimal execution with market orders, they could be used in the same way as the Almgren-Chriss model is used today, i.e., for scheduling. Important requirements are (i) to choose a (transient) market impact model that does not lead to price manipulation strategies, and (ii) to be able to estimate the parameters of the model in a way that is coherent with the execution process in the second layer of algorithms (see the appendix of this chapter). In 2015, practitioners do not seem to use these models for optimally scheduling execution processes. One of the main problems for practical use, even when the market impact model guarantees the absence of price manipulation, is that the optimal strategies obtained with this family of models are often unrealistic with large trades followed by quiet periods to wait for the price to decay.

7.1.3 Limit orders and dark pools

In the initial understanding of the Almgren-Chriss model, only market orders were used. It is therefore natural that a third strand of literature emerged, with execution models involving limit orders and/or the possibility to chase liquidity in dark pools.

---

[4] Another possibility is to consider

$$S_t = S_0 + \int_0^t f\left(\frac{v_s}{V_s}\right) G\left(\int_s^t V_u\,du\right) V_s\,ds + \sigma W_t. \tag{7.1}$$

The trade-off at the heart of the Almgren-Chriss model, and in all the models of the first two parts of the literature is between execution costs and price risk. In the third part of the literature, this trade-off still exists, but the execution costs can be reduced by using limit orders or dark pools (for instance mid-point dark pools). However, there is an additional risk entering the picture: the risk of not being executed.

The first optimal execution models with limit orders were the models of Bayraktar and Ludkovski [19], and Guéant, Lehalle, and Fernandez-Tapia [87]. In these models, the execution process only involves limit orders, that can be posted at any price. The probability for a given limit order to be hit (in fact, entirely filled, because there is no partial fill in these models) depends on the difference between the mid-price of the stock and the price of the limit order.

In the above models, and in some generalizations (see for instance [86][5]), only limit orders are allowed. In that case, the problem is solved using classical tools of stochastic optimal control. In particular, the optimal price at which orders should be posted is obtained through the resolution of a complex Hamilton-Jacobi-Bellman (HJB) equation. After these first models with limit orders only, some other models were built with both market orders and limit orders. The models proposed by Cartea and Jaimungal [41] and by Huitema [110] are interesting examples. In both cases, the trader can post limit orders at any price, and he can also take liquidity from the book using market orders. In these more general models, the optimal strategy is obtained by solving a quasi-variational inequality (QVI). In general, one cannot solve in closed-form the Hamilton-Jacobi-Bellman equation or the QVI. However, it is often possible to reduce the dimensionality of the problem to be able to solve it numerically.

These models have a lot of drawbacks: there is at most one limit order at a time, the current state of the limit order book is not taken into account, the size of the orders sent by the trader is often fixed, partial fills are seldom considered, the bid-ask spread is often assumed to be constant when market orders are allowed, etc. But one of the most important drawbacks is that limit orders can be posted at any price. In other words, the very discrete nature of the limit order book, with at most one queue of orders for each price that is a multiple of the tick size, is completely ignored. In practice, if we consider a standard stock, with a bid-ask spread of 2 ticks, a trader has to choose between a market order (or a marketable limit order), a limit order at the unique price between the best bid and the best ask (i.e., inside the spread), a limit order at the first limit, a limit order at the second limit, or a limit order deeper in the book. Therefore, the practical use of optimization models that make a difference between posting a limit order at 1.1 tick from the

---

[5]This article generalizes both [19] and [87]. It also discusses market making.

best bid and posting a limit order at 1.4 tick from the best bid is very questionable. Although it is often convenient to build models with a continuous state space to benefit from the power of differential calculus, it leads here to models that can hardly be used in practice, except maybe for small-tick stocks.

As we shall see in Chapter 11, execution models with limit orders (even when market orders are allowed) are very similar to market making models. Both kinds of models are always based on a simplified representation of the limit order book. The main difference is that execution models involve only buy orders or only sell orders, while in the case of market making, the trader buys and sells, and makes money out of the difference between its bid and ask quotes.[6] The models presented in [19] and [87] were both inspired by the market making model of Avellaneda and Stoikov [13].[7] In fact, many of the models proposed in the market making literature, or in the literature about high-frequency trading strategies, can be used to build models for optimal execution with both limit and market orders. Good models should take account of the discrete nature of the choice that has to be made by traders. A typical example of such a market making model that could be adapted to optimal execution issues is the market making model of Guilbaud and Pham [92].[8]

In addition to models for the execution of large orders with limit orders and market orders, a few papers have been written on optimal execution with market orders and orders sent to dark pools. Examples of such models are those of Kratz and Schöneborn [119, 120] or that of Horst and Naujokat [108] – see also [116], which deals with market impact models in the presence of dark pools.

All these papers are dealing with the optimal execution of an entire IS order. In other words, the models in this third part of the literature have been proposed to replace the model of Almgren and Chriss, not as a way to build algorithms and tactics following a given trading curve. However, as we will see in the next section, some of these models can be used to design tactics, even though they were not built for that purpose.

---

[6]Another difference is also that the initial portfolio, in the case of an optimal execution problem, contains the shares to be sold – if a long portfolio is to be liquidated – whereas it is usually empty in the case of a market making problem.

[7]We explain, in Chapter 11, why the model of Avellaneda and Stoikov is a bad model for the market making of most stocks, or more generally for market making in most order-driven markets, but why their modeling framework is adapted to market making in quote-driven markets.

[8]One could also cite the model of Hult and Kiessling [111], although it is not parsimonious enough to be used in practice.

7.2 Optimal execution models in practice

Among the numerous goals of the mathematicians – and academics from other fields – involved in Quantitative Finance, one is, or should be, to provide practitioners with new models and new tools for solving their quantitative problems. For instance, many models developed by academics in the area of derivatives pricing have definitely contributed to a better management of trading books comprised of vanilla and exotic derivatives products. Is it the case for quantitative issues in the cash equity world? The answer is not that obvious.

The academic literature on optimal execution is now very diverse, with different families of models, and a wide variety of academics involved in the field. However, as opposed to the literature on derivatives pricing, it is far less plug-and-play. Building execution algorithms is, in practice, far more complex than coding pricers, and the literature on optimal execution only addresses the quantitative/modeling part of the problem. Furthermore, even the quantitative part of the problem is only partially addressed, both because it is complex, and because most academics do not have a clear understanding of the practices in the brokerage industry and on the cash equity desks of corporate and investment banks.

7.2.1 The two-layer approach: strategy vs. tactics

As explained at the very beginning of Chapter 3, execution algorithms are usually made of two layers. The first layer is called the strategic layer. In short, it consists in computing an optimal trading curve, i.e., an ideal evolution of the portfolio, from its initial state to its complete liquidation. Then, the second layer, often referred to as the tactical layer, sends orders to one or several venues, in order to buy or sell shares, and follow the optimal trading curve.

The above description is a little simplistic, but it sheds light on the decomposition of the execution process into two different layers.

As far as the strategic layer is concerned, we have seen, in the previous chapters, how to compute optimal trading curves, for Implementation Shortfall, Target Close, VWAP, and TWAP orders.[9] Optimal trading curves are usu-

---

[9]The case of POV orders is different. We have developed in Chapter 4 a model to find the optimal rate of participation by assuming a deterministic volume curve. Then, in practice, one may directly use dynamic strategies to obtain a participation rate as close as possible to the desired one.

ally computed using pre-trade analytics (expected market volume, expected available liquidity, expected market impact, expected bid-ask spread, expected volatility, etc.). In fact, because parameters are usually estimated along with confidence bounds, several optimal trading curves can be computed, and the outcome of the strategic layer is not only a simple trading curve, but also upper and lower trading envelopes. For obtaining trading envelopes, another possibility is based on the optimal control function $(t, q) \mapsto v^*(t, q)$ which gives the optimal number of shares[10] that should be bought at time t when the portfolio contains q shares (at time t).[11] A common way to define envelopes is then as the upper and lower frontiers of the set of points defined by $\{(t, q) \in [0, T] \times \mathbb{R}, |v^*(t, q) - v^*(t, q^*(t))| \leq \alpha V_t\}$ for some constant α that sets the width between the two envelopes. In that case, envelopes are such that, at a time t, for a point (t, q) in the envelope, the optimal behavior is similar (up to a proportion α of the market volume) to what it would be on the optimal trajectory.

Once an optimal trading curve and the associated trading envelopes have been computed, the tactical layer of the algorithm is in charge of sending orders to the different venues in order to remain close to the optimal trading curve and, in all cases, inside the domain delimited by the upper and lower trading envelopes. The initial order (IS, TC, VWAP, etc.) is called the parent order. The smaller orders sent by the tactical layer to the different venues over the course of the execution process are called the child orders.

7.2.2 Child order placement

We have seen previously that the academic literature on optimal execution seldom deals with the optimal way to follow a given trading curve by sending orders of all kinds to lit and/or dark venues. However, a few dynamic models of the literature on optimal execution can be adapted to help making decisions within the framework of the tactical layer. Static models have also been proposed for the optimal placement of child orders. In the next paragraphs, we discuss the approaches based on static models. Then, we present an interesting dynamic model, and we discuss its applicability.

7.2.2.1 Static models for optimal child order placement

An optimal trading curve represents the (*ex-ante*) optimal scheduling of an execution process. Trading curves can be defined over horizons of a few minutes, dozen of minutes, or hours. In practice, optimal trading curves are often

---

[10]In fact an instantaneous volume, or a trading rate.

[11]This function v^* is for instance obtained by using the Hamilton-Jacobi equation associated with the Bolza problem at stake – see Eq. (8.6) for the case of an Implementation Shortfall order.

divided into short periods of time called slices. Several choices are possible for the slices. Slices with constant duration can be considered, with a duration corresponding to the horizon over which market impact and execution cost parameters have been estimated, or a duration adapted to the tactics used in the second layer. A variant is to consider slices with constant duration in volume time.[12] Slices of 1 to 5 minutes are common for liquid stocks when one uses a constant slice length, but it may go up to 10 or 15 minutes for illiquid stocks. Uneven slicing may also be used. For instance, slices may be computed so that the number of shares to be bought/sold over each slice would be constant if one followed the optimal trading curve computed in the first layer of the algorithm.

The tactical layer can use all kinds of orders and all types of venues in order to buy or sell shares. Over each slice, the goal assigned to the tactical layer is to buy or sell, at the best possible price, the number of shares that would permit to be on the optimal trading curve at the beginning of the next slice. Somehow, the problem at the level of a slice is the same as the global problem: optimally buying or selling a given number of shares. However, an important difference is that the tactical layer does not have the obligation to buy or sell the ideal number of shares: a target is set, but it may not be reached. In practice, in the case of the liquidation of a long portfolio, the tactics over a given slice may not have sold enough shares, for instance when a limit order has been posted and not filled. Conversely, the tactics may have sold too many shares compared to the ideal number, for instance if several limit orders posted on different venues have been hit almost simultaneously. Even though there is no obligation for the tactics to buy or sell an exact number of shares over each slice, the algorithm must ensure that the actual execution trajectory stays between the two trading envelopes discussed above. In particular, this requirement guarantees total liquidation by the final time T, because trading envelopes coincide with the optimal trading curve at time T.

Many authors have considered the problem of optimally buying or selling a small amount of shares over a short period of time. In the case of static models, everything works as if only one decision could be taken at the beginning of each slice. The basic trade-off of the literature is between market orders and limit orders.[13] In very simple models, such as those proposed in [118, 171, 176],

---

[12]This means that the expected market volume is the same over each slice.

[13]There are also a few papers dealing with iceberg orders and dark pools. For instance, Esser and Mönch [64] considered a very simple model to deal with the choice of the most important parameter of an iceberg order: the visible size – as a percentage of the total size. In particular, they focused on the trade-off between exposure and priority: a small visible size compared with the total size of the order guarantees low exposure, but a reduced probability to be executed (because of low priority), while a large one increases exposure, but increases the probability to be filled rapidly (because of higher priority). As far as dark pools are concerned, an interesting paper using some ideas of reinforcement learning is the paper of Laruelle, Lehalle, and Pagès [124], which provides a strategy to split an order across several dark pools.

there is only one venue and no partial fill. If the target number of shares to sell over a given slice is v, then, at the start of the slice (say at time $\underline{t}$), one can send a market order of size v, or post a limit order of size v at a price p_L. In the latter case, at the end of the slice (say at time $\bar{t}$), if the order has been filled – entirely filled in most models – then one gets the price p_L for each share. Otherwise, the shares are evaluated as if they were sold using a market order at time $\bar{t}$. In other words, if we denote by $\tau(p_L)$ the (random) time of fill for a limit order at price p_L, and by $p_M(\bar{t})$ the bid price (assuming there is enough liquidity at the first limit) at time $\bar{t}$, then the payoff per share is:

- $p_M(\underline{t})$ if a market order is sent at the start,

- $p_L 1_{\tau(p_L) \leq \bar{t}} + p_M(\bar{t}) 1_{\tau(p_L) > \bar{t}}$, if a limit order is posted at the start at price p_L.

The different models of this type proposed in the literature for computing the optimal behavior differ from one another in their assumptions regarding the law of $\tau(p_L)$ and $p_M(\bar{t})$, and in the different objective criterions considered by the authors (risk-neutral, mean-variance, etc.).

Models of this type are too simplistic for many reasons: they are static, there is usually only one venue in these models, partial fills are rarely considered, etc. Nevertheless, they are interesting because they highlight what is needed to address the trade-off between limit orders and market orders. The first estimation needed is an estimation of the probability to be filled at different price levels. With the above notations, it means an estimation of the law of $\tau(p_L)$ for various values of p_L. More generally, this can be a law for the number of shares that will be executed at each price. The second estimation needed is an estimation of the conditional distribution of $p_M(\bar{t})$, given that the limit order at price p_L has not been filled.

Within the family of static models, the model proposed by Cont and Kukanov [51] is certainly the only one that could be used in practice. It does not have indeed the drawbacks of most simple static models. In particular, partial fills are considered; instead of modeling the probability to be (entirely) filled at a given limit price, the focus is on the total outflow[14] at that price over a given period of time. Moreover, their model enables to take trading fees and rebates into account (see Chapter 2).

They consider a trader willing to buy v shares over a slice. In the single-venue version of their models, the algorithm can send, at time $\underline{t}$, a market order of size M and a limit order of size L at the first limit (the best bid). They assume that there is a transaction fee f for market orders and a rebate r

---

[14]The outflow takes into account both order cancellations and the arrival of liquidity-taking orders.

for limit orders. Obviously, limit orders are associated with a gain h and market orders with a cost h, where h is half the bid-ask spread. At time $\bar{t}$, they do not consider the random variable $p_M(\bar{t})$, but instead consider a penalty if the number of shares that have been bought is not equal to v.

To model the probability of a limit order to be hit, they consider the quantity V^b at the best bid before the order is sent and a random variable ξ standing for the outflow at the best bid. The number of shares bought at time $\bar{t}$ is then

$$M + (\xi - V^b)_+ - (\xi - V^b - L)_+.$$

The cost over the initial mid-price is

$$(h + f)M - (h + r)\left((\xi - V^b)_+ - (\xi - V^b - L)_+\right)$$

$$+\lambda\left(v - \left(M + (\xi - V^b)_+ - (\xi - V^b - L)_+\right)\right)_+,$$

where the last term stands for the cost (and the price of risk) for not having purchased v shares.[15]

Under technical assumptions on the values of the parameters, the authors show that there is an optimal[16] mix between market orders and limit orders, and they give the optimal expressions for L and M in closed-form. The formulas they obtain depend on the state of the order book, the fee structure, the order flow properties, and the aversion to non-execution risk (through λ).[17]

In addition to the fact that volumes are taken into account, one of the interest of [51] over the rest of the literature is that the approach is not bounded to the single-venue case. Cont and Kukanov also propose a similar model with multiple venues to post several limit orders (though still at the best bid price). In the multi-venue case, the framework is only slightly modified to account for the fact that one may post several limit orders for a total size larger than the target volume v: the objective function has to penalize the quantity obtained in excess of v. A numerical method is proposed in [51] to compute the optimal mix in the multi-venue case, because there is no closed-form solution to the problem beyond the single-venue case.

The model proposed in [51] has a lot of realistic features. An important limitation of the model (in addition to being a static model[18]) is however that the authors only consider limit orders posted at the first limit (and in fact at

---

[15]In the case of their single-venue model, there is no risk of buying too many shares, because the optimal strategy verifies $L + M = v$. This risk only comes with the possibility to buy shares over several venues.

[16]Optimality means that the expected value of the above cost is minimal.

[17]λ also proxies price risk, i.e., the risk associated with the volatility of the asset.

[18]One can always use a static model dynamically, by recomputing recurrently the optimal strategy to see if it is affected by new initial conditions.

the same price in the case of the multi-venue version of their model). Nevertheless, it is reasonable to think that the model could be generalized to allow for the choice of the price(s) at which to post limit orders.

7.2.2.2   Dynamic models for optimal child order placement

Most models proposed in the literature for child order placement are static models. However, a few dynamic execution models using limit orders and market orders (see Section 7.1), and some of the market making models proposed in the academic literature, can be adapted to design tactics. These dynamic models can usually be considered in two different manners: over each slice, or globally, over the whole execution process. The first possibility consists in considering that the problem over a slice is an optimal execution problem. Then, one can consider any relevant model of optimal execution with limit orders and market orders, and use it over a slice of a few minutes – usually after a relaxation of the terminal condition, because the number of shares to buy or sell over each slice constitutes a target rather than a constraint. The second possibility consists in using an optimal execution model over the whole time window $[0, T]$ of the execution process. In that case, the model has to be slightly modified, in order to penalize departure from the optimal trading curve computed *ex-ante* by the strategic layer of the algorithm.

In what follows, we consider a single-venue model for the liquidation of a long-only single-stock portfolio with q_0 shares. This model is freely inspired from the market making model of Guilbaud and Pham proposed in [92]. In addition to applying a similar modeling framework to optimal execution, we generalize the model proposed in [92] in numerous ways: more possible prices to post limit orders, additional variables to better assess the probabilities to be filled at the different limit prices, penalization terms to stay close to a pre-computed trading curve, etc. Our goal is to show what matters in the design of a dynamic model of optimal execution with limit orders and market orders, and, besides, to discuss the role of mathematics in the design of tactics.

We consider a simplified limit order book modeled by two processes: a process for the mid-price and a process for the bid-ask spread. The mid-price process $(S_t)_t$ is assumed to be a martingale with infinitesimal generator $\mathcal{L}$. The bid-ask spread $(\psi_t)_t$ is a continuous-time Markov chain with discrete state space $\{\delta, \ldots, J\delta\}$, where δ is the tick size. We denote by $(r^{\text{spread}}_{j\delta, j'\delta})_{j, j'}$ the instantaneous transition matrix of the process $(\psi_t)_t$.

In addition to the mid-price and the bid-ask spread, other variables may be used to describe the limit order book. In fact, our main goal is not to describe the limit order book, but rather to have a precise estimation of the probabilities to get executed for limit orders at different prices. Having good estimates of these probabilities is often the key to design efficient tactics. In practice, the probabilities to be executed at different prices depend on several variables or indicators: the bid-ask spread, the volumes V^a and V^b at the best ask price and at the best bid price respectively, the volume imbalance (i.e., the ratio $\frac{V^a-V^b}{V^a+V^b}$[19]), the order flow imbalance,[20] etc.

The bid-ask spread has already been introduced. Other indicators are regarded as a single variable (that may be multi-dimensional) denoted by I. We assume that $(I_t)_t$ is a continuous-time Markov chain with discrete state space $\{I_1,\ldots,I_K\}$.[21] We denote by $(r^{\text{ind}}_{k,k'})_{k,k'}$ the instantaneous transition matrix of the process $(I_t)_t$.[22]

We consider that limit orders all have the same size l. In particular, we write $l_t = l$ when a limit order is in the limit order book at time t and $l_t = 0$ otherwise.[23] They can be posted at the best ask, or at one tick from the best ask, either to improve the best ask price, or at the second limit to get a high priority in the queue. The execution of limit orders[24] is modeled by a point process $(N_t)_t$ with an intensity $\lambda(\psi_t, I_t, p_t)$ depending at time t on the bid-ask spread ψ_t, on the variable I_t, and on the price $S_t + \frac{\psi_t}{2} + p_t\delta$ of the limit order, where $p_t \in \{0,1\}$ if $\psi_t = \delta$, and $p_t \in \{-1,0,1\}$ otherwise.

In addition to limit orders, market orders can be sent.[25] We denote by J the jump process corresponding to these market orders, i.e., $J_t = J_{t-} + 1$ if a market order is sent at time t. We assume that the size of a market order is m_t (we choose this size in the interval $[0,\overline{m}]$), and that the execution price is $S_t - \frac{\psi_t}{2}$.[26]

The resulting dynamics for the number of shares in the portfolio is

$$dq_t = -l_t dN_t - m_t dJ_t.$$

---

[19]Ratios of the form $\frac{V^a-V^b}{V^a+V^b+H}$, where H proxies the hidden liquidity, can also be considered – see [12].

[20]See [40] for an optimal liquidation model in which the order flow is taken into account.

[21]Some variables, such as the volume imbalance, the order flow imbalance, or the volume at the first limits of the limit order book, are naturally defined as continuous variables. An important task in many approaches is to discretize these variables by finding the relevant ranges of value that indicate a significantly high or low probability to be executed at the different prices.

[22]Hereafter, we assume that the processes $(\psi_t)_t$ and $(I_t)_t$ are independent. It is straightforward to generalize the model, by considering a transition matrix for the process $(\psi_t, I_t)_t$.

[23]This assumption was not made in [92]. In fact, we assume a fixed size for the limit orders for two reasons: (i) for the sake of simplicity, and (ii) because the probability to be filled – in fact entirely filled – does not depend, in the model, on the size of the order. A consequence is that we also need to consider the case where it is optimal not to post any limit order.

[24]There is no partial fill in the model.

[25]The model can easily be generalized to take into account trading fees and rebates.

[26]$\overline{m}$ has to be small for this hypothesis to be relevant.

The associated dynamics for the cash account is given by

$$dX_t = l_t \left(S_t + \frac{\psi_t}{2} + p_t \delta \right) dN_t + m_t \left(S_t - \frac{\psi_t}{2} \right) dJ_t.$$

We now need to specify the optimization problem we intend to solve. We consider the use of this model over the time window $[0, T]$, which corresponds to the whole execution process. We also assume that an optimal trading curve (hereafter denoted by $t \mapsto q^*(t)$) has already been computed. Our goal is therefore to liquidate the portfolio while remaining close to this trading curve.

The objective function to maximize is of the form[27]

$$\mathbb{E} \left[X_T + q_T S_T - \ell(q_T) - \int_0^T g(q_t - q^*(t)) dt \right],$$

where the function ℓ penalizes incomplete execution, and the function g penalizes departure from the pre-computed optimal trading curve.[28]

To solve this problem, we introduce the value function $u(t, x, q, S, \psi, I)$. It is viscosity solution of the following QVI:

$$
\begin{aligned}
0 = \min \Big\{ &- \partial_t u(t, x, q, S, \psi, I) - \mathcal{L}u(t, x, q, S, \psi, I) + g(q - q^*(t)) \quad (7.2) \\
&- \sum_{j=1}^{J} r_{\psi, j\delta}^{\text{spread}} \left(u(t, x, q, S, j\delta, I) - u(t, x, q, S, \psi, I) \right) \\
&- \sum_{k=1}^{K} r_{I, I_k}^{\text{ind}} \left(u(t, x, q, S, \psi, I_k) - u(t, x, q, S, \psi, I) \right) \\
&- \left(\sup_{p \in Q_\psi} \lambda(\psi, I, p) \left(u \left(t, x + l \left(S + \frac{\psi}{2} + p\delta \right), q - l, S, \psi, I \right) \right. \right. \\
&\qquad\qquad\qquad\qquad \left. \left. - u(t, x, q, S, \psi, I) \right) \right)_+; \\
&u(t, x, q, S, \psi, I) - \sup_{m \in [0, \overline{m}]} u \left(t, x + m \left(S - \frac{\psi}{2} \right), q - m, S, \psi, I \right) \Big\},
\end{aligned}
$$

where $Q_\psi = \{0, 1\}$ if $\psi = \delta$, and $Q_\psi = \{-1, 0, 1\}$ otherwise.

---

[27]Unlike in the previous chapters, we do not consider an expected utility framework with a CARA utility function. Instead, we consider a risk neutral objective function, because the price risk has already been considered in the trading curve q^*. The risk of not being executed is also somehow taken into account in the penalization terms.

[28]The same ideas could be used over a slice. In that case, there is less need to penalize departure from the trading curve q^*.

The terminal condition is

$$u(T, x, q, S, \psi, I) = x + qS - \ell(q).$$

In order to compute u and the optimal strategy, we write u as $u(t, x, q, S, \psi, I) = x + qS + \phi_{\psi,I}(t, q)$. It is straightforward to verify that $(\phi_{\psi,I})_{\psi,I}$ is viscosity solution of

$$
\begin{aligned}
0 = \min\Bigg\{ &-\partial_t \phi_{\psi,I}(t, q) + g(q - q^*(t)) - \sum_{j=1}^{J} r_{\psi,j\delta}^{\text{spread}} (\phi_{j\delta,I}(t, q) - \phi_{\psi,I}(t, q)) \\
&- \sum_{k=1}^{K} r_{I,I_k}^{\text{ind}} (\phi_{\psi,I_k}(t, q) - \phi_{\psi,I}(t, q)) \\
&- \left(\sup_{p \in Q_\psi} \lambda(\psi, I, p) \left(l \left(\frac{\psi}{2} + p\delta \right) + \phi_{\psi,I}(t, q - l) - \phi_{\psi,I}(t, q) \right) \right)_+ ; \\
&\phi_{\psi,I}(t, q) - \sup_{m \in [0, \overline{m}]} \left(-m \frac{\psi}{2} + \phi_{\psi,I}(t, q - m) \right) \Bigg\},
\end{aligned}
\tag{7.3}
$$

with terminal condition $\phi_{\psi,I}(T, q) = -\ell(q)$.

These equations deserve several remarks. First, the complex quasi-variational inequality (7.2) involves a function of 6 variables. By using $(\phi_{\psi,I})_{\psi,I}$, the problem boils down to the 2D system of equations (7.3) involving functions of 2 variables. The solution of this system can very easily be approximated numerically on a grid – see [92]. Subsequently, the optimal action in every state can be calculated and tabulated in advance. It depends on the state of both the execution process through the couple (t, q), and the limit order book – or more generally the trading environment – through the couple (ψ, I). Second, $(\phi_{\psi,I})_{\psi,I}$ does not depend on $\mathcal{L}$. In other words, the solution of the problem is independent of the price dynamics, as long as the price process is a martingale. In particular, it is independent of the volatility. This is not surprising, given the objective criterion we have considered. Moreover, the volatility has already been taken into account in the computation of q^*, and the function ℓ may also depend on volatility.

One could think that such a model could be used to design tactics. However, to be used in practice, it must account for the presence of several venues. In particular, there should be as many bid-ask spread processes as there are lit venues – and these processes are obviously not independent from one another. Moreover, the state space for the variable I must be enlarged in order to account for the different signals provided by the state of the different limit order books in the different lit venues. Taking into account the toxicity of the different venues also matters. Eventually, adding constraints in order to stay inside the trading envelopes is required. Even though these additional features

136   *The Financial Mathematics of Market Liquidity*

do not raise any problem theoretically, a model with all these features would require computer-intensive calculations.

One of the requirements for using such a model is to have precise estimates of the probabilities to be filled at the relevant limit prices, for each value of the bid-ask spread(s) and the other indicator(s) I. These probabilities can be estimated on high-frequency data and/or by using a market simulator.[29] A different approach, with the same state space, consists in using a learning approach for finding the optimal behavior in each state, without relying on pre-computed probabilities to be executed. Nevmyvaka et al. proposed such a machine learning approach in [149]. More generally, applications of machine learning – especially reinforcement learning techniques – to optimal execution issues may be one of the next "hot topics" for academic research on optimal execution.

7.2.2.3 Final remarks on child order placement

Most of the academic models are too simple to be used in practice. Adding more realistic features to the models proposed in the literature is possible, but the resulting models then turn out to be very demanding in terms of computation time. Furthermore, realistic stochastic optimal control models are naturally based on very large state spaces. Therefore, estimating with precision the probabilities to be executed, conditionally on each state, may be complicated, unless one uses large datasets.[30] A solution may be to simplify the state space[31] in order to reduce its size, and eventually obtain satisfying estimates for the probabilities to be filled.

In addition to the large datasets often required to calibrate realistic models, and the long computation time required to get the optimal action in each state of these models, a key issue with the academic models on optimal execution is that the management of limit orders in limit order books is never addressed. In all models, at most one limit order (per venue) is considered at a time, whereas in practice it is often a better idea to post several limit orders at different prices in the limit order book, in order to have a high priority in the queues. Nevertheless, tactics involving several limit orders in the same limit order book at the same time can be built by using (in an unusual manner) a stochastic optimal control model such as the one previously described. The idea is to anticipate the next best action, i.e., the action that will be optimal once the posted limit order is filled. This enables to post a second order – and the same reasoning applies recursively to post as many orders as

---

[29]In all cases, some assumptions need to be made to deal with partial fills in the estimation procedure.

[30]Estimation cannot be done with data over long periods of time, because of non-stationarity (this is also a problem with learning approaches – see [149]).

[31]Clustering techniques can be used.

desired. However, the question of the optimal number of limit orders to post is not addressed. Posting a lot of limit orders enables to have a high priority in the different queues, but it presents two drawbacks: (i) information leakage, because posting an order reveals a willingness to buy or sell, and (ii) some latency risk, because one may not have the time to cancel orders, should the market move (this issue is particulary important for VWAP orders).

We have previously described several academic models that we find relevant. However, it must be noted that no single mathematical model can provide "the" optimal tactic. In particular, trading opportunities cannot all come from the solution of an HJB equation or a QVI. In practice, rule-based tactics are often used, that are not based on complex stochastic optimal control, but on basic intuitions. In front of so many possibilities, like in the current economic environment, one could be tempted to give up, to leave mathematics aside, and to rely on simple trading rules. However, we believe that mathematics can definitely help designing tactics. The optimization and stochastic optimal control methods used in this book are well suited, but machine learning techniques should also be considered.

7.3   Conclusion

This chapter closes Part II on optimal liquidation. In the first chapters, we have presented in detail the framework proposed by Almgren and Chriss. Models *à la* Almgren-Chriss can be used in numerous situations to optimally schedule the execution of a large order: different types of execution, single-stock and multi-asset portfolios, etc. In this chapter, we have discussed the different models proposed in the academic literature beyond the models inspired by the Almgren-Chriss framework. We have discussed some models with limit orders and some models with dark pools. In particular, we have shown that some of them are adapted to build tactics aimed at buying or selling shares, while respecting the pace imposed by a given trading curve (pre-computed, for instance, to mitigate execution costs and price risk).

We hope that this chapter will help readers develop some new tactics. We also hope that readers working in academia will develop some new approaches, more specifically some new methods based on machine learning techniques.

Appendix to Chapter 7: Market impact estimation

In order to compute trading curves and trading envelopes, one needs pre-trade analytics. Among the statistics needed, the key ones are related to market impact and execution costs. We now briefly review the numerous approaches that have been proposed to model the market impact of trades. We also insist on the fact that market impact estimates based on execution data – not only market data – should be used for building trading curves.

Many approaches exist to estimate market impact, and it is noteworthy that not only the methodologies are different across papers, but the very definition of market impact differs from one study to the next.

For instance, some authors study the market impact of single trades, while some others consider the market impact associated with entire execution processes. Some simply want to estimate market impact as a function of the volume traded, while some others want to both estimate market impact and understand the price formation process (these are two intimately related problems – see Bouchaud et al. [31]). Some look at the immediate impact of single trades (comparison between the price before the trade and the price before the next trade), while some others look at the temporal evolution of market impact (for single trades or for long sequences of trades). And so on.

In terms of data, market data, such as Trade and Quote (TAQ) data, is often used. Some interesting research works are also based on data from exchanges. The most relevant estimates for optimal execution purposes are based on execution data from brokers, from the cash trading desks of investment banks, or from hedge funds with in-house execution. However, for obvious reasons, only a few results based on execution databases have been published.

Historically, the literature on market impact estimation was divided into two parts.

A first set of articles deals with the market impact of single trades. Researchers studied the evolution of prices following a buyer-initiated trade or a seller-initiated trade,[32] and the influence of the volume traded on the magnitude of the price impact. Many papers on market impact are devoted to single trades. As a consequence, the literature covers various markets and different periods of time. Lillo et al. [136, 137] studied the immediate impact of more

---

[32]The information that a transaction is triggered by a buyer or a seller may not be present in common datasets. However, it is often possible to infer this information from the data. The algorithm commonly used is due to Lee and Ready, see [125].

than 100 million trades from the 1990s using TAQ data on the New York Stock Exchange. Hopman [107] studied the market impact of single trades as a function of the volume transacted on French stocks 30 minutes after the trades (the dataset was made of transactions from the 1990s). Other examples include [67] and [68], in which the immediate impact of a single trade is modeled as a function of the trade size for American and British stocks. The temporal evolution of market impact is studied for three French and British stocks in the paper [154] by Potters and Bouchaud, using a few months of data from 2002. In [27], Bladon et al. used data from the Spanish Stock Exchange about the individual activity of members over the period 2001–2004 for estimating the immediate impact of single trades.

Be it for studying the immediate impact of a trade, or the temporal evolution of the market price following a trade, the approaches used in this (early) part of the literature are questionable in their basic forms, especially if they were to be applied on recent datasets. Most authors find a concave shape for the influence of the volume, but there is nowadays little information in the volume of a single trade, because the size of a single trade is mainly driven by the available liquidity in limit order books.[33] Furthermore, when one goes beyond the immediate impact and measures the impact of a trade after a given period of time, it is very difficult to know, with market data only, whether the impact comes from that specific trade, or from the following trades. In particular, the law of large numbers cannot be invoked because the time series of trade sides/signs are positively autocorrelated: buyer-initiated trades (respectively seller-initiated trades) tend to be followed by buyer-initiated trades (respectively seller-initiated trades) – see Bouchaud et al. [31], Lillo and Farmer [135], or Tóth et al. [168].

A second kind of approach for measuring market impact consists in relating volumes to price changes by using aggregated data. The basic idea is that there should not be any price change (on average) if the volume of buyer-initiated trades were equal to the volume of seller-initiated trades. Subsequently, a way to estimate market impact is to consider, over time windows of a given length, the difference between the volume/number of buyer-initiated trades and the volume/number of buyer-initiated trades, and use it as an explanatory variable for the price change over the same period.[34]

Several articles are based on this idea of using trade imbalance as a proxy for extra volume (compared to an equilibrium situation), in order to estimate the market impact as a function of the volume traded. For instance, Plerou et

---

[33]Nowadays, almost no liquidity-consuming order reaches the second limit of a limit order book.

[34]Often, the relevant variable is the proportion of seller-initiated trades, or the difference between the volume/number of buyer-initiated trades and the volume/number of seller-initiated trades, normalized by the total volume/number of trades.

al. [153], who used data from the 1990s on 116 US stocks, estimated that the price change is a concave power function of the volume. In particular, they found different exponents for the power functions, depending on the length of the time window over which measures were carried out. Weber and Rosenow [172] used data from 2002 for the ten most frequently traded stocks on the NASDAQ, and measured both trade imbalances and price changes over time windows of 5 minutes. They fitted a power function with an exponent 0.76.

The approaches based on aggregated data over bins of several minutes are often used because of their simplicity.[35] However, by using aggregated data, one focuses more on the elasticity of supply and demand than on market impact. In particular, trade imbalances and price changes are considered over the same period of time, therefore there is no dynamic consideration. Actually, most of the papers in this strand of the literature on market impact estimation do not study which proportion of the impact is only temporary, at what speed it decays, etc.

Over the last ten years, the literature on market impact has developed in new directions.

First, several theories have been developed to understand market impact. Measuring market impact is one thing, but understanding why market impact has a specific form is certainly more important from a scientific viewpoint. A good theory for the market impact of trades was actually the missing link to reconcile the long-range autocorrelation of order flow documented in the literature, and the celebrated efficient market hypothesis which states (in particular) that price changes cannot be predicted (i.e., price processes should be martingales). Not surprisingly, this literature was mainly initiated by econophysicists, i.e., by researchers who tend to see a market or an exchange as a physical system which reacts to trades in a way that has to be coherent with some predefined fundamental laws. Several very interesting articles are found in this literature: Bouchaud et al. [31, 32], Farmer et al. [65, 66] – see also the recent paper of Donier [59].

Second, the literature on market impact has gone beyond the impact of liquidity-consuming orders. In the early literature, the market impact was implicitly the reaction of prices to a market order, or at least a marketable limit order. Limit order posting and cancellation also have an influence on the book:

---

[35]For someone who only has access to market data, using a method of this kind is certainly the best option to estimate market impact. However, some assumptions need to be made if one wants to attribute price changes to the execution cost component or to the permanent impact component of a model *à la* Almgren-Chriss. Moreover, the length of the time windows should be chosen so as to obtain stable results, and in line with the slicing of the trading curve.

this was studied recently, for instance in Cont et al. [52], or in Eisler et al. [63].

Third, a new literature emerged that focuses on the impact of large orders, or metaorders. In particular, this literature focuses on the impact of the execution process of parent orders, and not on the impact of single trades – or at least single trades/child orders are not considered independently from their large parent orders. The focus is on the evolution of the price over the course of the whole execution process of the parent order/metaorder, and after the end of the execution process. Two kinds of datasets are used for studying the market impact of metaorders: (i) datasets coming from stock exchanges, with, for each trade, the identity of the trader, and (ii) execution datasets coming from brokerage companies, cash trading desks at investment banks, or large asset managers having their own execution algorithms. In the former case, parent orders need to be "implicited" from the data: trades are packed into coherent sequences of trades to form plausible parent orders. In the latter case, the decomposition of parent orders into child orders is documented.

The first study in that direction[36] is the one of Almgren et al. [10] who used execution data from Citigroup to estimate the parameters needed in the Almgren-Chriss model. With the notations of Chapter 3, it means that they estimated the value of k and the form of the function L.

Almgren et al. [10] considered a set of around 700,000 metaorders. For each metaorder, they computed the slippage and the impact of the execution process on the price.[37] In the Almgren-Chriss model, with the notations of Chapter 3, these two quantities are respectively

$$\frac{X_T - X_0 - q_0 S_0}{q_0 S_0} = -\frac{k}{2S_0} q_0 - \frac{1}{q_0 S_0} \int_0^T L\left(\frac{v_t}{V_t}\right) V_t dt + \frac{\sigma}{S_0} \int_0^T \frac{q_t}{q_0} dW_t, \quad (7.4)$$

and

$$S_T - S_0 = kq_0 + \sigma W_T. \quad (7.5)$$

---

[36]Some old studies do exist actually. For instance, Chan and Lakonishok [48] used a dataset comprised of the trades of 37 large investment management firms from mid-1986 to the end of 1988, in NYSE and AMEX stocks. They used some simple rules to reconstitute parent orders, and then studied market impact. The market impact of large orders that are not split into child orders was also studied in the old paper of Keim and Madhavan [115]. They considered around 4,000 transactions on the upstairs market in the United States between 1985 and 1990, and studied the respective impact of OTC and non-OTC transactions. More recently, Carollo et al. [36] studied the difference in price impact between trades executed on the SETS (LSE) and trades executed off-market.

[37]In [10], the authors consider the market price a long time after the end of the execution process, in order to measure the permanent component of market impact. However, for optimal execution problems, we are not really interested in what happens long after the end of the execution process. Here, we consider the price at the end of the execution process. What is called permanent market impact in the Almgren-Chriss model is actually not really permanent, because the price impact (partially) decays after the end of the execution process.

Eq. (7.5) makes it possible to estimate k. In fact, k is a function of S_0 and also of the liquidity of the stock, that can be measured by the average daily volume or the market capitalization of the stock. The point here is that Eq. (7.5) enables to estimate the change in price due to the execution process with basic estimation techniques. Almgren et al. [10] used nonlinear estimation techniques, because they did not consider a linear function of q_0, but instead a power function of q_0. Empirically, they found an exponent around 0.9, i.e., almost in line with the linear assumption. However, there is a consensus among researchers around smaller values of the exponent, around 0.5, and most often between 0.4 and 0.7: this is the so called square-root law of market impact.[38]

To estimate L, the approach consists in assuming a constant market volume curve V, and a constant participation rate $\rho = \frac{q_0}{VT}$ for each of the metaorders.[39] Then, we combine Eqs. (7.4) and (7.5) to obtain

$$\frac{X_T - X_0 - q_0 \frac{S_0 + S_T}{2}}{q_0 S_0} = -\frac{1}{S_0} \frac{L(\rho)}{\rho} + \epsilon,$$

where the residual is $\epsilon = \frac{\sigma}{S_0} \left(\int_0^T \left(1 - \frac{t}{T}\right) dW_t - \frac{1}{2} W_T \right)$.

By using nonlinear techniques, one can estimate the value of the parameters η, ϕ, ψ in the case of an execution cost function $L(\rho) = \eta |\rho|^\phi + \psi |\rho|$, or in the case of a more complex model, because η and ψ are not constant, but instead functions of the price S_0, the volatility of the stock, the average bid-ask spread, etc. To better fit the data, it is also possible to make η and ψ functions of q_0 or T, although it is then harder to justify the model (and some results do not hold anymore, such as those related to POV orders).

We refer the interested reader to [10] for more details on the estimations of the parameters of a model *à la* Almgren-Chriss using this approach.[40] The fundamental point with this method is that the execution cost function and the market impact function estimated here are related to the tactics of Citigroup. This is one of the most essential points when it comes to estimating market impact and execution costs for optimal execution purposes: what needs to be measured is not the reaction of the market to orders in general, but in-

---

[38]The square-root law of market impact applies to the relation between the price change $S_T - S_0$ and the volume traded q_0 (appropriately renormalized), i.e., $S_T - S_0 \propto \pm |q_0|^{\frac{1}{2}}$. A similar relation does not necessarily hold for the part of the impact that remains after the post-execution decay.

[39]This assumption is particularly relevant for POV orders, VWAP orders, and TWAP orders. In particular, datasets need to be filtered to only keep the parent orders for which execution was carried out at an approximately constant pace.

[40]The methodology can be improved by removing components of the price changes that could be explained by non-idiosyncratic factors – for instance using the CAPM. Estimations can also be improved by better taking account of heteroscedasticity.

stead the reaction of the market to the way one trades and the average costs associated with the tactics used in the second layer of execution algorithms. In particular, the better the tactics in the second layer of execution algorithms, the lower the value of the execution cost function used in the first (strategic) layer for computing optimal trading curves.

Other authors have used execution data on metaorders to study market impact. For instance, Moro et al. [148] used data from the Spanish Stock Market (2001–2004) and from the London Stock Exchange (2002–2004), Tóth et al. [167] used a proprietary database of metaorders on futures markets, from Capital Fund Management, over the period 2007–2010, Bershova and Rakhlin [20] used a proprietary dataset of large trades executed between 2009 and 2011 by AllianceBernstein for around 50 institutional funds, Brokmann et al. [34] used another dataset from Capital Fund Management with 1.6 million metaorders, over the period 2011–2013, on equity markets (Europe, United States, Japan, and Australia), etc. These studies do not aim at estimating the parameters of a model which will be used for optimal execution purposes, but at studying the evolution of the price over the course of the execution process, and after the end of the execution process. In particular, their results show a square-root dependence on the traded volume (normalized by the average daily volume) of the difference in price between the beginning and the end of the execution process: this is the square-root law of market impact. These studies also exhibit a concave evolution of the price change over the course of the execution process, suggesting more impact at the beginning of the execution of a metaorder than at the end. There is however no consensus on the decay after the end of the execution process, and on the really permanent impact of trades.

This literature raises many questions about the appropriate normalization factor for the price changes and for the volume traded. It also raises questions about the nature of the metaorders. In particular, the metaorders of a hedge fund can hardly be compared to the metaorders of institutional investors going through a brokerage company. An important topic is also the duration of the trade, or equivalently the role of the participation rate. In particular, the latter is at the heart of the market impact model used in Almgren-Chriss, but it is often absent from the above studies. In fact, because we consider a deterministic market volume curve, we may consider V_t as a renormalization factor across stocks and periods of times, instead of regarding $\frac{v_t}{V_t}$ as the true participation rate at time t. The estimations obtained for the whole process can then be rescaled to estimate the execution cost functions L.

Recently, two papers have discussed the role of the participation rate in the market impact of metaorders. Bacry et al. [15] for instance used a dataset of 400,000 metaorders from Crédit Agricole Cheuvreux, and confirmed the square-root law of market impact, but only for groups of trades with similar durations. Interestingly, they identified an effect of the duration T in $1/T^{0.25}$. In other words, the lower the participation rate of a trade, the lower the market impact, *ceteris paribus.* Zarinelli et al. [177] also analyzed the role of participation rates, but the nature of the large database of metaorders they used makes their results questionable.

The literature on market impact is very diverse, and the main objective for researchers in this field is definitely not to estimate the permanent market impact parameter[41] nor the execution cost function of models *à la* Almgren-Chriss. Nevertheless, it is possible to estimate the parameters needed to build an optimal scheduling curve by using methods similar to those used in this literature on market impact estimation.[42] In particular, some important ideas should be retained from the literature on market impact, such as the use of volatility or bid-ask spread as a pre-factor for price changes, the fact that the market component of price changes should be factored out, etc. The most important one is certainly that large databases are necessary to obtain stable estimations. Ideally, for building an optimal execution model, the estimation of the market impact and execution cost model should be based on execution data of metaorders, because the goal is to evaluate the quality of tactics. However, if one does not have a large set of execution data, then it may be better to use aggregated market data for proxying executed volumes by trade imbalances, and then estimating the market price reaction (as explained above).

---

[41] It is important to recall that the permanent market impact in the Almgren-Chriss model has little to do with the permanent market impact after the decay of the temporary one. It represents instead a part of the slippage that cannot be avoided, because it is independent from the strategy employed. It is however not the only one, because the total cost associated with the linear component of L is also independent of the strategy – at least for buy-only and sell-only strategies.

[42] This is not limited to models *à la* Almgren-Chriss. In particular, the square-root law suggests to choose a transient market impact model where, in Eq. (7.1), f and G are power law functions with respective exponents 0.5 and −0.5.

Part III

Liquidity in Pricing Models

Chapter 8

Block trade pricing

A great deal of my work is just playing with equations and seeing what they give.

— Paul A. M. Dirac

The literature on optimal execution deals with the optimal way to build or unwind a position, that is, with execution strategies and tactics. In this chapter, we go beyond strategies and tactics, and we show how the models of the previous chapters – with a focus on the Almgren-Chriss framework – can be used to give a price to a block of shares. This is important to price risk trades, but also more generally to give a price to (il)liquidity. In this chapter, we introduce in particular the concept of risk-liquidity premium that can be used for pricing purposes, but also to penalize illiquid portfolios in many models.

8.1 Introduction

What is the market price of a stock? At a given point in time, it can be defined as the mid-price of the stock, or the price of the last trade on the primary venue. Other definitions can be considered, for instance to account for what happens on other venues (think of the NBBO[1] in the United States). In all cases, the market price, or Mark-to-Market (MtM) price,[2] reflects or approximates the price at which one could buy or sell a few shares. Although market prices are often used to give a price to (or evaluate) large portfolios, all practitioners know that it would be impossible to buy or sell on the market a large number of shares at their current MtM price.

MtM prices are often criticized in accountability for many reasons: the inadequacy between investment horizon and continuous or daily update of prices, the mismatch between the evaluation of assets and the evaluation of

---

[1]National Best Bid and Offer.
[2]Sometimes, this price is also called the fair value.

liabilities (think of an insurance company), the too high variability of MtM prices, etc. Supporters of fair value accounting often argue, however, that the MtM price is the most accurate evaluation of an asset, because it corresponds to a recent transaction or to a price close to the prices at which agents propose to buy or sell. Many criticisms are possible but one of the most unquestionable ones is that MtM prices do not reflect the liquidity of assets. Two stocks, or more generally two assets, can indeed have the same MtM price, but if the first asset is more liquid than the second one, then buying the same large quantity of both assets costs more in the case of the second asset than in the case of the first one.

In this chapter, we leverage the models presented in Chapters 3 and 4 in order to introduce an alternative to MtM pricing for large portfolios of stocks. The main idea behind our approach is that the value of a block of shares (here a long position) should depend on both the costs and the risk of liquidating the block of shares on the market. We focus on the maximum price at which an agent with no inventory would be ready to buy a block of shares. This price, which is nothing but the indifference price (see Appendix A) of the lottery consisting in getting the shares and liquidating them using the optimal liquidation strategy, will be called the price of the block, or the block trade price (with a slight abuse of language, because it is a value rather than a price). It takes account of the MtM price of the portfolio, of the execution costs, and of the market impact, but also of the risk associated with the execution process to unwind the position. Because the MtM price remains the benchmark price, we introduce the concept of risk-liquidity premium, which is defined as the difference between the block trade price and the associated MtM price of the same block of shares.

Risk-liquidity premia are important to give a price to blocks of shares in the case of risk trades. In Part II of this book, we have mainly dealt with agency trades, in which brokers or cash traders execute orders on behalf of clients, using the strategy chosen by the client. Another possibility for financial intermediaries is to propose firm quotes to the clients,[3] and then to execute orders at their own discretion. In that case, we speak of risk trades or principal trades. In the case of agency trades, the risk is borne by the client, while in the case of risk trades, the financial intermediary has acquired the position and unwinds it at his own risk.[4] In practice, as of today, risk trades are largely

---

[3] A firm quote can be a cash amount or a spread over a benchmark price. This benchmark price can be the MtM price, but it can also be a price that is unknown at the time of the deal. This is the case with Guaranteed VWAP contracts in which the client is guaranteed to get the VWAP over a given period plus or minus a fixed spread/fee (decided upon at the start).

[4] In the case of risk trades, the identity of the client is important. Through the analysis of past transactions, one can often see that some clients are better informed than others. In practice, if the buy (sell) orders of a specific client are often followed by an increase (decrease) in the stock price, then orders coming from this client ought to be executed

subsidized (although not explicitly) by agency trades. In effect, agents pay a fee to use the algorithms of brokers and electronic desks of investment banks, and risk trades are often executed at MtM price unless the size of the order is really too large.[5] This situation is economically inefficient, and it might come to an end with MiFID 2. In particular, if the status of Systematic Internalisers is adopted by many,[6] many market participants will need to provide bid and offer prices for different sizes.

Risk-liquidity premia are also important outside of execution issues. A risk-liquidity premium is indeed nothing but a penalty that applies to illiquid portfolios. As soon as one wants to take liquidity into account in a model, risk-liquidity premia can be useful. In Chapter 9, for instance, we use risk-liquidity premia in a pricing model for options on illiquid assets. Other applications include, for example, asset management and portfolio choice.

In this chapter, we mainly focus on risk-liquidity premia in the case of single-stock portfolios, as we can obtain closed-form formulas in this case. Nevertheless, we start with the multi-asset case, in order to provide a general definition. In the last section of this chapter, we also cover the case of Guaranteed VWAP contracts, in which the risk-liquidity premium is measured with respect to the VWAP, and not with respect to the MtM price.

8.2 General definition of block trade prices and risk-liquidity premia

8.2.1 A first definition

Let us consider an agent with a portfolio with positions in d stocks. We denote by $q_0 = (q_0^1, \ldots, q_0^d)$ the vector of positions. We assume that the agent wants to sell his portfolio, and asks a financial intermediary for a price.[7] The problem we consider is that of the intermediary: pricing the portfolio.

To solve this problem, we assume that the intermediary has no initial position, and that he is going to liquidate the portfolio on the market at his

---

faster. In practice, this may be done by increasing the risk aversion parameter γ, or by introducing a drift in the price dynamics – see Chapter 5.

[5]If a specific agent uses a too large proportion of risk trades, he may also be charged.

[6]Contrary to what happened with the first version of MiFID.

[7]We do not only consider the case of a long-only portfolio: the positions can be positive or nonpositive. In particular, in the single-asset case, if q_0 is negative, then it means that the financial intermediary sells shares to the agent, and the price given by the intermediary to the agent is thus negative.

own risk once he buys it. To start, we can assume that he will liquidate the portfolio by using an unconstrained strategy over a time interval $[0, T]$. The strategy used by the intermediary is then naturally an Implementation Short-fall strategy.

If the intermediary buys the portfolio at time 0 for a price P, then, using the notations of the Almgren-Chriss framework (see Chapters 3 and 5), his expected utility at time 0 is

$$\sup_{(v_t)_t \in \mathcal{A}} \mathbb{E}\left[-\exp(-\gamma(X_T - P))\right], \tag{8.1}$$

where[8]

$$
\begin{aligned}
dq_t^i &= v_t^i dt, \forall i, \\
dS_t^i &= \sigma^i dW_t^i + k^i v_t^i dt, \forall i, \\
dX_t &= \sum_{i=1}^{d} -v_t^i S_t^i dt - V_t^i L^i \left(\frac{v_t^i}{V_t^i}\right) dt,
\end{aligned}
$$

and

$$\mathcal{A} = \left\{ (v_t)_{t \in [0,T]} \in \mathbb{H}^0(\mathbb{R}^d, (\mathcal{F}_t)_t), \forall i, \int_0^T v_t^i dt = -q_0^i, \int_0^T |v_t^i| dt \in L^\infty(\Omega) \right\}.$$

If the intermediary does not buy the portfolio, then his utility is

$$-\exp(-\gamma X_0). \tag{8.2}$$

The indifference price $P = P_T(q_0)$ that equalizes Eq. (8.1) and Eq. (8.2) is given by[9]

$$P_T(q_0) = \sup_{(v_t)_t \in \mathcal{A}} -\frac{1}{\gamma} \log\left(\mathbb{E}\left[\exp(-\gamma(X_T - X_0))\right]\right).$$

Using Eq. (5.6), and all the analysis of Section 5.3.2, we get

$$P_T(q_0) = \sum_{i=1}^{d} q_0^i S_0^i - \sum_{i=1}^{d} \frac{k^i}{2} q_0^{i\,2} - \min_{q \in \mathcal{C}_T} J_T(q), \tag{8.3}$$

where[10]

$$J_T(q) = \int_0^T \left(\sum_{i=1}^{d} V_t^i L^i \left(\frac{q^{i'}(t)}{V_t^i}\right) + \frac{\gamma}{2} q(t) \cdot \Sigma q(t) \right) dt,$$

---

[8]We assume that the d-dimensional process $(\sigma^1 W_t^1, \ldots, \sigma^d W_t^d)_t$ has a nonsingular co-variance matrix Σ.

[9]See also Definition A.3 of Appendix A.

[10]$q(t)$ is always a column vector in what follows.

and
$$\mathcal{C}_T = \left\{ q \in W^{1,1}(0,T), q(0) = q_0, q(T) = 0 \right\}.$$

This indifference price $P_T(q_0)$ deserves several remarks. First, it is made of two components:

$$P_T(q_0) = \underbrace{\sum_{i=1}^{d} q_0^i S_0^i}_{\text{MtM price}} - \underbrace{\left(\sum_{i=1}^{d} \frac{k^i}{2} q_0^{i\,2} + \min_{q \in \mathcal{C}_T} J_T(q) \right)}_{\text{risk-liquidity premium}}.$$

The first component of the indifference price $P_T(q_0)$ is the MtM price of the portfolio. The second part is the risk-liquidity premium to be subtracted from the MtM price in order to get the indifference price. The risk-liquidity premium represents the amount that the agent needs to pay to compensate for (i) the market impact and execution costs incurred when liquidating the portfolio, and (ii) the risk that the price moves over the course of the execution process.

A second important remark on the indifference price $P_T(q_0)$ is that it depends on T. But what is T? In the previous chapters, we considered execution processes over a given period of time because an agency order was sent to be executed over a pre-defined time window (except in the case of POV orders). However, in the case of a risk trade, there is no natural terminal time.[11] In fact, we can consider the limit $T \to +\infty$.

8.2.2 A time-independent definition

To make the definition of the block trade price of a portfolio independent of T, we first state a result in the form of a lemma:

Lemma 8.1. *The function $T \in \mathbb{R}_+^* \mapsto \min_{q \in \mathcal{C}_T} J_T(q)$ is nonincreasing.*

Proof. *Let us consider $T < T'$, and a function $q \in \mathcal{C}_T$. Let us define the function $\tilde{q} \in \mathcal{C}_{T'}$ by $\tilde{q}(t) = 1_{t \in [0,T]} q(t)$. We have*

$$\min_{\mathcal{C}_{T'}} J_{T'} \leq J_{T'}(\tilde{q}) = J_T(q).$$

This being true for all $q \in \mathcal{C}_T$, we have $\min_{\mathcal{C}_{T'}} J_{T'} \leq \min_{\mathcal{C}_T} J_T$.

Because J_T is a nonnegative function, we know from Lemma 8.1 that $\lim_{T \to +\infty} \min_{q \in \mathcal{C}_T} J_T(q)$ does exist.

We are now ready for the following definition:

Definition 8.1. *The block trade price of a portfolio with positions q_0 is defined by*

$$P(q_0) = \sum_{i=1}^{d} q_0^i S_0^i - \sum_{i=1}^{d} \frac{k^i}{2} q_0^{i\,2} - \lim_{T \to +\infty} \min_{q \in \mathcal{C}_T} J_T(q). \qquad (8.4)$$

---

[11]Sometimes, the end of the trading day is a natural terminal time.

The risk-liquidity premium associated with this portfolio is

$$\ell(q_0) = \sum_{i=1}^{d} \frac{k^i}{2} q_0^{i\,2} + \lim_{T \to +\infty} \min_{q \in \mathcal{C}_T} J_T(q). \tag{8.5}$$

This definition raises at least two sets of questions. First, sending T to $+\infty$ is an interesting theoretical idea, but what does it mean in practice? If it means an execution process over several days, how can we handle it? Second, how can we compute, in practice, a block trade price, or equivalently a risk-liquidity premium?

With respect to the first question, it is noteworthy that, although it was not built to deal with execution processes lasting more than a few hours, the Almgren-Chriss framework can in effect be used over a few days. The market volume curve $(V_t)_t$ can indeed cover several days, and be set to 0 when the market is closed. When the market is closed, the market price continues to diffuse, but the model prevents the trader from trading. Therefore, sending T toward $+\infty$ is not a real issue as far as modeling is concerned. Furthermore, since $P(q_0)$ is an upper bound for all the prices $P_T(q_0)$, the block trade price gives the maximum price at which the intermediary could accept to buy the portfolio from the agent.

With respect to the second question, we can approximate $P(q_0)$ using the methods of Chapter 6. The approach consists in (i) choosing $\tilde{T}$ sufficiently large, (ii) finding the minimizer q^* of $J_{\tilde{T}}$ by using one of the algorithms of Chapter 6, and (iii) evaluating $\lim_{T \to +\infty} \min_{q \in \mathcal{C}_T} J_T(q)$ by $J_{\tilde{T}}(q^*)$. Then, the block trade price of the portfolio can be approximated by considering

$$P(q_0) \simeq \sum_{i=1}^{d} q_0^i S_0^i - \sum_{i=1}^{d} \frac{k^i}{2} q_0^{i\,2} - J_{\tilde{T}}(q^*).$$

The goal in the next section is to show that $\lim_{T \to +\infty} \min_{q \in \mathcal{C}_T} J_T(q)$ can in fact be evaluated directly, without relying on the previous approximations, at least in the case of a single-stock portfolio.

8.3   The specific case of single-stock portfolios

We have seen above that computing a risk-liquidity premium boils down to computing $\lim_{T \to +\infty} \min_{q \in \mathcal{C}_T} J_T(q)$. In the case of single-asset portfolios, this limit can be found in closed-form, as soon as we assume that the market volume curve is flat.[12] The goal of this section is to show how to obtain the formulas for the risk-liquidity premium and the block trade price. Although the formula is rather simple, the way to obtain it is technical, because it relies on the theory of viscosity solutions for first-order Hamilton-Jacobi partial differential equations. The reader may choose to skip the next (technical) paragraphs, and directly go to Section 8.3.2, in which closed-form formulas are presented.

8.3.1 The value function and its asymptotic behavior

We consider the case of a single-stock portfolio. We also assume that the market volume curve is flat, with $V_t = V$. We define the value function θ_T associated with the Bolza problem exhibited in Chapter 3 by

$$\theta_T : (t, \mathfrak{q}) \in [0, T) \times \mathbb{R} \mapsto \min_{q \in \mathcal{C}_{t,T,\mathfrak{q}}} \int_t^T \left(VL\left(\frac{q'(s)}{V}\right) + \frac{1}{2}\gamma\sigma^2 q^2(s) \right) ds,$$

where

$$\mathcal{C}_{t,T,\mathfrak{q}} = \left\{ q \in W^{1,1}(t, T), q(t) = \mathfrak{q}, q(T) = 0 \right\}.$$

$\mathfrak{q}S_t - \frac{k}{2}\mathfrak{q}^2 - \theta_T(t, \mathfrak{q})$ represents the indifference price at which a trader would be ready to buy $\mathfrak{q}$ shares at time t, if the shares have to be liquidated on the market before time T. We are specifically interested in the asymptotic behavior of $\theta_T(0, q_0)$ when $T \to +\infty$, but we first study the function θ_T.

A first straightforward remark is that, because we have assumed a flat market volume curve, $\theta_T(t, \cdot)$ only depends on the difference between T and t. This point plays an important role in the determination of the asymptotic behavior.

Let us start with a lemma.

Lemma 8.2. *The following properties hold:*

1. $\forall(t, \mathfrak{q}) \in \mathbb{R}_+ \times \mathbb{R}, T \in (t, +\infty) \mapsto \theta_T(t, \mathfrak{q})$ *is a nonincreasing function.*

2. $\forall \mathfrak{q} \in \mathbb{R}, \forall T \in \mathbb{R}_+^*, t \in [0, T) \mapsto \theta_T(t, \mathfrak{q})$ *is a nondecreasing function.*

3. $\forall T \in \mathbb{R}_+, \forall t \in [0, T), \mathfrak{q} \in \mathbb{R} \mapsto \theta_T(t, \mathfrak{q})$ *is a convex function.*

4. $\forall T \in \mathbb{R}_+, \forall t \in [0, T), \mathfrak{q} \in \mathbb{R} \mapsto \theta_T(t, \mathfrak{q})$ *is a nonincreasing function on $\mathbb{R}_-$ and a nondecreasing function on $\mathbb{R}_+$.*

---

[12]Assuming a flat market volume curve is only a real issue if the size of the portfolio is so large that a significant part of the execution process still needs to be executed at the end of the first day.

Proof. *The proof of the first point is the same as for Lemma 8.1.*

The second point can be deduced from the first one. Let us consider a couple $(t_1, t_2) \in [0, T)^2$, *with* $t_1 < t_2$. *We have*

$$\theta_T(t_1, \cdot) = \theta_{T+t_2-t_1}(t_2, \cdot) \leq \theta_T(t_2, \cdot).$$

The third point is more classical. Let us consider $\mathfrak{q} \in \mathbb{R}$ *and* $h > 0$. *Let us denote by* $q_{t,\mathfrak{q}-h} \in \mathcal{C}_{t,T,\mathfrak{q}-h}$ *and* $q_{t,\mathfrak{q}+h} \in \mathcal{C}_{t,T,\mathfrak{q}+h}$ *the respective minimizers of*

$$\int_t^T \left(VL\left(\frac{q'(s)}{V}\right) + \frac{1}{2}\gamma\sigma^2 q^2(s) \right) ds$$

over $\mathcal{C}_{t,T,\mathfrak{q}-h}$ *and* $\mathcal{C}_{t,T,\mathfrak{q}+h}$.

Let us define $q = \frac{1}{2}q_{t,\mathfrak{q}-h} + \frac{1}{2}q_{t,\mathfrak{q}+h} \in \mathcal{C}_{t,T,\mathfrak{q}}$. *We have, by definition of* θ_T *and convexity of* L,

$$\theta_T(t, \mathfrak{q} + h) - 2\theta_T(t, \mathfrak{q}) + \theta_T(t, \mathfrak{q} - h)$$

$$\geq \int_t^T V\left(L\left(\frac{q'_{t,\mathfrak{q}+h}(s)}{V}\right) - 2L\left(\frac{q'(s)}{V}\right) + L\left(\frac{q'_{t,\mathfrak{q}-h}(s)}{V}\right) \right) ds$$

$$+ \frac{1}{2}\gamma\sigma^2 \int_t^T \left(q_{t,\mathfrak{q}+h}(s)^2 - 2q(s)^2 + q_{t,\mathfrak{q}-h}(s)^2 \right) ds$$

$$\geq 0.$$

Therefore, $\theta_T(t, \cdot)$ *is a convex function.*

For the fourth point, we prove the result on $\mathbb{R}_+$ *(the same reasoning would apply on* $\mathbb{R}_-$*).*

Let us consider $(\mathfrak{q}_1, \mathfrak{q}_2) \in \mathbb{R}_+^2$ *with* $\mathfrak{q}_1 < \mathfrak{q}_2$. *Let us denote by* $q_{t,\mathfrak{q}_2} \in \mathcal{C}_{t,T,\mathfrak{q}_2}$ *the minimizer of*

$$\int_t^T \left(VL\left(\frac{q'(s)}{V}\right) + \frac{1}{2}\gamma\sigma^2 q^2(s) \right) ds$$

over $\mathcal{C}_{t,T,\mathfrak{q}_2}$.

We have, using the monotonicity properties of L,

$$\theta_T(t, \mathfrak{q}_1) \leq \int_t^T \left(VL\left(\frac{\mathfrak{q}_1}{\mathfrak{q}_2}\frac{q'_{t,\mathfrak{q}_2}(s)}{V}\right) + \frac{1}{2}\gamma\sigma^2 \left(\frac{\mathfrak{q}_1}{\mathfrak{q}_2}q_{t,\mathfrak{q}_2}(s)\right)^2 \right) ds$$

$$\leq \int_t^T \left(VL\left(\frac{q'_{t,\mathfrak{q}_2}(s)}{V}\right) + \frac{1}{2}\gamma\sigma^2 q_{t,\mathfrak{q}_2}(s)^2 \right) ds$$

$$= \theta_T(t, \mathfrak{q}_2).$$

Therefore, $\theta_T(t, \cdot)$ *is a nondecreasing function on* $\mathbb{R}_+$.

Using Lemma 8.2, we can state the asymptotic behavior of the sequence of functions $(\theta_T)_T$.

Proposition 8.1. $\forall (t, \mathfrak{q}) \in \mathbb{R}_+ \times \mathbb{R}$, $\theta_T(t, \mathfrak{q})$ *converges, when T tends to $+\infty$, towards a value $\theta_\infty(\mathfrak{q})$ independent of t.*

Furthermore, the convergence of $(\theta_T)_T$ towards the convex function θ_∞ is locally uniform on $\mathbb{R}_+ \times \mathbb{R}$.

Proof. *By the first point of Lemma 8.2, $(\theta_T)_T$ is a nonincreasing sequence of nonnegative functions. Therefore, $\lim_{T \to +\infty} \theta_T(t, \mathfrak{q})$ exists.*

For $(t_1, t_2) \in [0, T)^2$, if $t_1 < t_2$, then we have $\theta_T(t_1, \cdot) = \theta_{T+t_2-t_1}(t_2, \cdot)$.

Therefore, $\lim_{T \to +\infty} \theta_T(t, \mathfrak{q})$ is in fact independent of t, and we define

$$\theta_\infty : \mathfrak{q} \in \mathbb{R} \mapsto \lim_{T \to +\infty} \theta_T(t, \mathfrak{q}).$$

Using the third point of Lemma 8.2, θ_∞ is a convex function and it is therefore continuous.

By Dini's theorem, the convergence of the sequence of convex (and therefore continuous) functions $(\theta_T(t, \cdot))_T$ towards θ_∞ (for a fixed $t \in \mathbb{R}_+$) is locally uniform on $\mathbb{R}$.

Using the second point of Lemma 8.2, we have that the convergence of $(\theta_T)_T$ towards θ_∞ is in fact locally uniform on $\mathbb{R}_+ \times \mathbb{R}$.

We now want to write θ_∞ in closed-form. For that purpose, we use the theory of viscosity solutions for first-order Hamilton-Jacobi partial differential equations.

Proposition 8.2. θ_∞ *is a viscosity solution on $\mathbb{R}$ of the first-order Hamilton-Jacobi equation:*

$$-\frac{1}{2}\gamma\sigma^2 q^2 + VH(-\theta'_\infty(q)) = 0,$$

with

$$\theta_\infty(0) = 0.$$

In particular, because θ_∞ is convex, the equation holds in the classical sense almost everywhere.

Proof. *Using classical arguments of the theory of viscosity solutions (see [16]) we can prove that[13] θ_T is a locally Lipschitz function, and is a viscosity solution of the first-order Hamilton-Jacobi equation:*

$$-\partial_t \theta_T(t, q) - \frac{1}{2}\gamma\sigma^2 q^2 + VH(-\partial_q \theta_T(t, q)) = 0, \qquad \text{on } [0, T) \times \mathbb{R}. \qquad (8.6)$$

---

[13] See [82] for more details on the proof.

Now, using the locally uniform convergence of $(\theta_T)_T$ toward θ_∞ obtained in Proposition 8.1, and a classical stability result for first-order Hamilton-Jacobi equations (see for instance Proposition 2.2 in [16]), we obtain that θ_∞ is a viscosity solution on $\mathbb{R}$ of the first-order Hamilton-Jacobi equation:

$$-\frac{1}{2}\gamma\sigma^2 q^2 + VH(-\theta'_\infty(q)) = 0, \qquad \text{on } \mathbb{R}.$$

Eventually, we have $\theta_T(t,0) = 0, \forall t \in [0,T)$, and therefore $\theta_\infty(0) = 0$.

Proposition 8.2 uses deep results that may not be known by the reader. The understanding of the proof is not important for what follows. However, the differential characterization of θ_∞ is fundamental to obtain the closed-form formula stated in the following theorem.

Theorem 8.1. *Assume that L is such that its Legendre-Fenchel transform H is decreasing on $\mathbb{R}_-$ and increasing on $\mathbb{R}_+$. Then,*

$$\theta_\infty(q) = \begin{cases} -\int_0^q H_-^{-1}\left(\frac{\gamma\sigma^2}{2V}x^2\right) dx & \text{if } q \geq 0, \\ -\int_0^q H_+^{-1}\left(\frac{\gamma\sigma^2}{2V}x^2\right) dx & \text{if } q < 0, \end{cases}$$

where H_+^{-1} and H_-^{-1} are respectively the inverses of the restriction to $\mathbb{R}_+$ and $\mathbb{R}_-$ of the function H.

Proof. *By using Proposition 8.2, we know that almost everywhere on $\mathbb{R}$ we have the following equality:*

$$H(-\theta'_\infty(q)) = \frac{\gamma\sigma^2}{2V}q^2.$$

By the fourth point of Lemma 8.2, we know that θ_∞ is nonincreasing on $\mathbb{R}_-$ and nondecreasing on $\mathbb{R}_+$. Therefore, we can use the two functions H_+^{-1} and H_-^{-1} to write (almost everywhere, but in fact everywhere)

$$\theta'_\infty(q) = \begin{cases} -H_-^{-1}\left(\frac{\gamma\sigma^2}{2V}q^2\right) & \text{if } q \geq 0, \\ -H_+^{-1}\left(\frac{\gamma\sigma^2}{2V}q^2\right) & \text{if } q < 0. \end{cases}$$

Using the fact that $\theta_\infty(0) = 0$, we have

$$\theta_\infty(q) = \begin{cases} -\int_0^q H_-^{-1}\left(\frac{\gamma\sigma^2}{2V}x^2\right) dx & \text{if } q \geq 0, \\ -\int_0^q H_+^{-1}\left(\frac{\gamma\sigma^2}{2V}x^2\right) dx & \text{if } q < 0. \end{cases}$$

8.3.2 Closed-form formula for block trade prices

Using Definition 8.1 and the results obtained in the previous section, we know that

$$
\begin{aligned}
P(q_0) &= q_0 S_0 - \frac{k}{2} q_0^2 - \lim_{T \to +\infty} \min_{q \in \mathcal{C}_T} J_T(q) \\
&= q_0 S_0 - \frac{k}{2} q_0^2 - \lim_{T \to +\infty} \theta_T(0, q_0) \\
&= q_0 S_0 - \frac{k}{2} q_0^2 - \theta_\infty(q_0).
\end{aligned}
$$

Using Theorem 8.1, we can therefore state the main theorem of this chapter:

Theorem 8.2. *Assume that the market volume curve is flat with $V_t = V$. Assume also that L is such that its Legendre-Fenchel transform H is decreasing on $\mathbb{R}_-$ and increasing on $\mathbb{R}_+$.*

If $q_0 \geq 0$, then the block trade price of a portfolio with q_0 shares is

$$
P(q_0) = q_0 S_0 - \frac{k}{2} q_0^2 + \int_0^{q_0} H_-^{-1}\left(\frac{\gamma \sigma^2}{2V} x^2\right) dx,
$$

where H_-^{-1} is the inverse of the restriction to $\mathbb{R}_-$ of H.

The risk-liquidity premium is therefore

$$
\ell(q_0) = \frac{k}{2} q_0^2 - \int_0^{q_0} H_-^{-1}\left(\frac{\gamma \sigma^2}{2V} x^2\right) dx.
$$

If $q_0 < 0$, then the block trade price of a portfolio with q_0 shares is

$$
P(q_0) = q_0 S_0 - \frac{k}{2} q_0^2 + \int_0^{q_0} H_+^{-1}\left(\frac{\gamma \sigma^2}{2V} x^2\right) dx,
$$

where H_+^{-1} is the inverse of the restriction to $\mathbb{R}_+$ of H.

The risk-liquidity premium is therefore

$$
\ell(q_0) = \frac{k}{2} q_0^2 - \int_0^{q_0} H_+^{-1}\left(\frac{\gamma \sigma^2}{2V} x^2\right) dx.
$$

Theorem 8.2 provides a formula to compute the risk-liquidity premium, or equivalently the block trade price. The formulation is a bit complicated because it covers all forms of functions L. In the case of an even function L, the expression can be simplified because H is even and therefore $H_-^{-1} = -H_+^{-1}$.

Corollary 8.1. *Assume that the market volume curve is flat with $V_t = V$. Assume also that L is an even function, such that its Legendre-Fenchel transform H is increasing on $\mathbb{R}_+$.*

The block trade price of a portfolio with q_0 shares is

$$P(q_0) = q_0 S_0 - \frac{k}{2}q_0^2 - \int_0^{|q_0|} H_+^{-1}\left(\frac{\gamma\sigma^2}{2V}x^2\right) dx, \tag{8.7}$$

and the risk-liquidity premium is

$$\ell(q_0) = \frac{k}{2}q_0^2 + \int_0^{|q_0|} H_+^{-1}\left(\frac{\gamma\sigma^2}{2V}x^2\right) dx. \tag{8.8}$$

Corollary 8.1 gives a simple expression but one must be able to compute H_+^{-1} from L, and then compute the integral in Eqs. (8.7) and (8.8).

In the special case where $L(\rho) = \eta|\rho|^{1+\phi}$, we know that

$$H(p) = \phi\eta\left(\frac{|p|}{\eta(1+\phi)}\right)^{1+\frac{1}{\phi}}.$$

Therefore,

$$H_+^{-1}(x) = \eta(1+\phi)\left(\frac{x}{\eta\phi}\right)^{\frac{\phi}{1+\phi}},$$

and

$$\int_0^{|q_0|} H_+^{-1}\left(\frac{\gamma\sigma^2}{2V}x^2\right) dx = \frac{\eta^{\frac{1}{1+\phi}}}{\phi^{\frac{\phi}{1+\phi}}}\frac{(1+\phi)^2}{1+3\phi}\left(\frac{\gamma\sigma^2}{2V}\right)^{\frac{\phi}{1+\phi}}|q_0|^{\frac{1+3\phi}{1+\phi}}.$$

Therefore, we can write the block trade price and the risk-liquidity premium in closed-form.

Corollary 8.2. *Let us suppose that $V_t = V$ and that $L(\rho) = \eta|\rho|^{1+\phi}$. Then, we have*

$$P(q_0) = q_0 S_0 - \frac{k}{2}q_0^2 - \frac{\eta^{\frac{1}{1+\phi}}}{\phi^{\frac{\phi}{1+\phi}}}\frac{(1+\phi)^2}{1+3\phi}\left(\frac{\gamma\sigma^2}{2V}\right)^{\frac{\phi}{1+\phi}}|q_0|^{\frac{1+3\phi}{1+\phi}},$$

and

$$\ell(q_0) = \frac{k}{2}q_0^2 + \frac{\eta^{\frac{1}{1+\phi}}}{\phi^{\frac{\phi}{1+\phi}}}\frac{(1+\phi)^2}{1+3\phi}\left(\frac{\gamma\sigma^2}{2V}\right)^{\frac{\phi}{1+\phi}}|q_0|^{\frac{1+3\phi}{1+\phi}}.$$

If $L(\rho) = \eta|\rho|^{1+\phi} + \psi|\rho|$, then $H(p) = 0$, for $|p| \leq \psi$. Therefore, Theorem 8.2 does not apply because the hypothesis of strict monotonicity of H on $\mathbb{R}_+$ is violated. However, the additional term $\psi|\rho|$ simply adds a total cost $\psi|q_0|$ to the execution. Therefore, we can still obtain closed-form formulas.

Corollary 8.3. *Let us suppose that $V_t = V$ and that $L(\rho) = \eta|\rho|^{1+\phi} + \psi|\rho|$. Then, we have*

$$P(q_0) = q_0 S_0 - \frac{k}{2}q_0^2 - \psi|q_0| - \frac{\eta^{\frac{1}{1+\phi}}}{\phi^{\frac{\phi}{1+\phi}}} \frac{(1+\phi)^2}{1+3\phi} \left(\frac{\gamma\sigma^2}{2V}\right)^{\frac{\phi}{1+\phi}} |q_0|^{\frac{1+3\phi}{1+\phi}},$$

and

$$\ell(q_0) = \frac{k}{2}q_0^2 + \psi|q_0| + \frac{\eta^{\frac{1}{1+\phi}}}{\phi^{\frac{\phi}{1+\phi}}} \frac{(1+\phi)^2}{1+3\phi} \left(\frac{\gamma\sigma^2}{2V}\right)^{\frac{\phi}{1+\phi}} |q_0|^{\frac{1+3\phi}{1+\phi}}.$$

Corollaries 8.2 and 8.3 provide closed-form formulas. In particular, we see that the risk-liquidity premium is not linear in q_0, but instead strictly convex in q_0, because of both the permanent component (the term in q_0^2) and the instantaneous component (the term in $|q_0|^{\frac{1+3\phi}{1+\phi}}$) of market impact.

The formulas we have obtained also make it possible to carry out some comparative statics to better understand the role played by the different parameters.

With respect to the risk parameters:

- The higher σ, that is, the more volatile the stock price, the higher the risk-liquidity premium.

- The higher γ, that is, the more risk adverse the trader, the higher the risk-liquidity premium.

With respect to the cost parameters:

- The higher k, that is, the higher the permanent market impact effect, the higher the risk-liquidity premium. This is natural as the price decreases over the course of a selling process and increases over the course of a buying process.

- The higher the proportional fee ψ, that is, the wider the bid-ask spread or the larger the stamp duty for instance, the higher the risk-liquidity premium.

- The higher η, that is, the higher the execution costs, the higher the risk-liquidity premium. In other words, the less liquid the stock, the higher the risk-liquidity premium.

- Similarly, the higher the market volume V, the lower the risk-liquidity premium.

8.3.3 Examples and discussion

The above formulas make it possible to compute very rapidly the value of the risk-liquidity premium associated with a block trade. In Table 8.1, we provide a few examples of risk-liquidity premia (in relative terms) for the same parameters as in Chapter 6, and for different values of the block size q_0 and risk aversion parameter γ.

We distinguish in particular the component (I) coming from permanent market impact, from the other component (II) related to temporary market impact and risk:

$$\ell(q_0) = \underbrace{\frac{k}{2}q_0^2}_{\text{(I)}} + \underbrace{\psi|q_0| + \frac{\eta^{\frac{1}{1+\phi}}}{\phi^{\frac{\phi}{1+\phi}}}\frac{(1+\phi)^2}{1+3\phi}\left(\frac{\gamma\sigma^2}{2V}\right)^{\frac{\phi}{1+\phi}}|q_0|^{\frac{1+3\phi}{1+\phi}}}_{\text{(II)}}.$$

The parameters for the stock (inspired from the stock TOTAL SA) are the following:

- $S_0 = 45$ €,

- $\sigma = 0.6$ €·day$^{-1/2}$·share^{-1}, i.e., an annual volatility approximately equal to 21%,

- $V = 4{,}000{,}000$ shares·day^{-1},

- $L(\rho) = 0.1|\rho|^{0.75} + 0.004|\rho|$,

- $k = 2 \times 10^{-7}$.

TABLE 8.1: Risk-liquidity premia for different values of the parameters.

γ (€$^{-1}$)	q_0	$\frac{\ell(q_0)}{q_0 S_0}$ (basis points)		
		(I)	(II)	Total
10^{-6}	200,000	4.4	5.1	9.5
5.10^{-6}	200,000	4.4	9.3	13.7
10^{-5}	200,000	4.4	12.2	16.6
10^{-6}	400,000	8.8	8.6	17.4
5.10^{-6}	400,000	8.8	16.2	25.0
10^{-5}	400,000	8.8	21.5	30.3

We see that the value of the risk-liquidity premium for a given block of shares is highly influenced by the choice of the risk aversion parameter γ. Therefore, the formulas of Theorem 8.2 and its corollaries are useful to be consistent across deals. Another important use of the above closed-form formulas is for the choice of γ. We have seen in Section 4.3.3 that γ could be implicited from the optimal choice of a participation rate. Here it can be implicited from the premium a trader would charge for a risk trade. One can indeed ask a cash trader for the prices he would quote for different stocks and different trade sizes, and estimate from his answers a value for γ.

8.3.4 A straightforward extension

Definition 8.1 can be generalized in a straightforward manner to include a maximum constraint on the participation rate, as in Section 5.2. It is indeed reasonable to assume that the portfolio bought by the trader in a risk trade will be liquidated progressively, with a cap $\rho_{\max}$ on the participation rate to the market. In that case, the block trade price is defined as

$$P_{\rho_{\max}}(q_0) = q_0 S_0 - \frac{k}{2} q_0^2 - \lim_{T \to +\infty} \min_{q \in \mathcal{C}_{\rho_{\max},T}} J_T(q),$$

and the associated risk-liquidity premium is defined as

$$\ell_{\rho_{\max}}(q_0) = \frac{k}{2} q_0^2 + \lim_{T \to +\infty} \min_{q \in \mathcal{C}_{\rho_{\max},T}} J_T(q),$$

where

$$\mathcal{C}_{\rho_{\max},T} = \left\{ q \in W^{1,1}(0,T), q(0) = q_0, q(T) = 0, |q'(t)| \le \rho_{\max}, \text{a.s. on } (0,T) \right\}.$$

Using the same reasoning as above, and the notations of Section 5.2, we obtain a result similar to Theorem 8.2:

Theorem 8.3. *Assume that the market volume curve is flat with $V_t = V$. Assume also that L is such that $H_{\rho_{max}}$ is decreasing on $\mathbb{R}_-$ and increasing on $\mathbb{R}_+$.*

If $q_0 \ge 0$, then the block trade price of a portfolio with q_0 shares is

$$P_{\rho_{max}}(q_0) = q_0 S_0 - \frac{k}{2} q_0^2 - \int_0^{q_0} H_{\rho_{max},+}^{-1} \left(\frac{\gamma \sigma^2}{2V} x^2 \right) dx,$$

where $H_{\rho_{max},+}^{-1}$ is the inverse of the restriction to $\mathbb{R}_+$ of $H_{\rho_{max}}$.

The risk-liquidity premium is therefore

$$\ell_{\rho_{max}}(q_0) = \frac{k}{2} q_0^2 + \int_0^{q_0} H_{\rho_{max},+}^{-1} \left(\frac{\gamma \sigma^2}{2V} x^2 \right) dx.$$

If $q_0 < 0$, then the block trade price of a portfolio with q_0 shares is

$$P_{\rho_{max}}(q_0) = q_0 S_0 - \frac{k}{2} q_0^2 - \int_0^{q_0} H_{\rho_{max},-}^{-1} \left(\frac{\gamma \sigma^2}{2V} x^2 \right) dx,$$

where $H_{\rho_{max},-}^{-1}$ is the inverse of the restriction to $\mathbb{R}_-$ of $H_{\rho_{max}}$.

The risk-liquidity premium is therefore

$$\ell_{\rho_{max}}(q_0) = \frac{k}{2} q_0^2 + \int_0^{q_0} H_{\rho_{max},-}^{-1} \left(\frac{\gamma \sigma^2}{2V} x^2 \right) dx.$$

8.4 A simpler case with POV liquidation

In the above paragraphs, we have dealt with block trade prices and risk-liquidity premia in the case of the liquidation of a portfolio with an Implementation Shortfall (IS) strategy. In this section we deal with the case of liquidation using a POV strategy. Because the POV strategy is more constrained than the IS strategy, it is clear that risk-liquidity premia will be higher for POV strategies than for IS strategies, all else equal. The question is how much higher? Is there a real difference between an IS strategy and a POV strategy in terms of risk-liquidity premium?

When a trader liquidates a single-stock portfolio at constant participation rate ρ, the liquidation process terminates at the first time T such that $\rho \int_0^T V_t dt = q_0$. We have seen in Chapter 4 (see Eq. (4.9)) that the amount on the cash account at time T is

$$X_T = X_0 + q_0 S_0 - \frac{k}{2} q_0^2 + \sigma \rho \int_0^T \int_t^T V_s ds dW_t - \frac{L(\rho)}{\rho} q_0.$$

Therefore, we can define a new notion of block trade price by

$$P_{\text{POV}}(q_0) = q_0 S_0 - \frac{k}{2} q_0^2 - \min_{\rho > 0} \left(\frac{L(\rho)}{\rho} q_0 + \frac{\gamma}{2} \sigma^2 \rho^2 \int_0^T \left(\int_t^T V_s ds \right)^2 dt \right),$$

and the associated risk-liquidity premium is defined as

$$\ell_{\text{POV}}(q_0) = \frac{k}{2} q_0^2 + \min_{\rho > 0} \left(\frac{L(\rho)}{\rho} q_0 + \frac{\gamma}{2} \sigma^2 \rho^2 \int_0^T \left(\int_t^T V_s ds \right)^2 dt \right).$$

Using Theorem 4.2, we can find, in the case of a flat market volume curve, closed-form formulas for the block trade price and the risk-liquidity premium.

Theorem 8.4. *Let us suppose that $V_t = V$ and that $L(\rho) = \eta|\rho|^{1+\phi} + \psi|\rho|$. Then, we have (for $q_0 > 0$)*

$$P_{POV}(q_0) = q_0 S_0 - \frac{k}{2} q_0^2 - \psi q_0 - (1+\phi) \eta^{\frac{1}{1+\phi}} \left(\frac{\gamma \sigma^2}{6\phi V} \right)^{\frac{\phi}{1+\phi}} q_0^{\frac{1+3\phi}{1+\phi}},$$

and

$$\ell_{POV}(q_0) = \frac{k}{2} q_0^2 + \psi q_0 + (1+\phi) \eta^{\frac{1}{1+\phi}} \left(\frac{\gamma \sigma^2}{6\phi V} \right)^{\frac{\phi}{1+\phi}} q_0^{\frac{1+3\phi}{1+\phi}}.$$

Proof. *We know from Theorem 4.2, that*

$$\frac{L(\rho)}{\rho}q_0 + \frac{\gamma}{2}\sigma^2\rho^2 \int_0^T \left(\int_t^T V_s ds \right)^2 dt$$

$$= \psi q_0 + \eta\rho^\phi q_0 + \frac{\gamma}{2}\sigma^2\rho^2 V^2 \frac{T^3}{3}$$

$$= \psi q_0 + \eta\rho^\phi q_0 + \frac{\gamma}{2}\sigma^2 \frac{q_0^3}{3\rho V}$$

is minimized for

$$\rho = \left(\frac{\gamma\sigma^2}{6\eta\phi} \frac{q_0^2}{V} \right)^{\frac{1}{1+\phi}}.$$

Therefore,

$$\ell_{POV}(q_0) = \frac{k}{2}q_0^2 + \psi q_0 + \eta \left(\frac{\gamma\sigma^2}{6\eta\phi} \frac{q_0^2}{V} \right)^{\frac{\phi}{1+\phi}} q_0 + \gamma\sigma^2 \frac{q_0^3}{6V} \left(\frac{\gamma\sigma^2}{6\eta\phi} \frac{q_0^2}{V} \right)^{-\frac{1}{1+\phi}}$$

$$= \frac{k}{2}q_0^2 + \psi q_0 + (1+\phi)\eta^{\frac{1}{1+\phi}} \left(\frac{\gamma\sigma^2}{6\phi V} \right)^{\frac{\phi}{1+\phi}} q_0^{\frac{1+3\phi}{1+\phi}}.$$

The formula for the risk-liquidity premium in the case of a liquidation at constant participation rate

$$\ell_{\text{POV}}(q_0) = \frac{k}{2}q_0^2 + \psi q_0 + (1+\phi)\eta^{\frac{1}{1+\phi}} \left(\frac{\gamma\sigma^2}{6\phi V} \right)^{\frac{\phi}{1+\phi}} q_0^{\frac{1+3\phi}{1+\phi}}$$

is very similar in its form to the formula for the risk-liquidity premium in the case of a liquidation with an IS strategy:

$$\ell(q_0) = \frac{k}{2}q_0^2 + \psi q_0 + \frac{\eta^{\frac{1}{1+\phi}}}{\phi^{\frac{\phi}{1+\phi}}} \frac{(1+\phi)^2}{1+3\phi} \left(\frac{\gamma\sigma^2}{2V} \right)^{\frac{\phi}{1+\phi}} q_0^{\frac{1+3\phi}{1+\phi}}.$$

Not surprisingly, the only difference is in the term linked to the optimization of the liquidation process, and $\ell(q_0) \le \ell_{\text{POV}}(q_0)$. What is more interesting is that the ratio $\frac{\ell(q_0)}{\ell_{POV}(q_0)}$ can be bounded from below independently of the value of the parameters.

Proposition 8.3.

$$\frac{\ell(q_0)}{\ell_{POV}(q_0)} \ge \frac{e\log(3)}{2\sqrt{3}}(\ge 0.86).$$

Proof.

$$
\begin{aligned}
\frac{\ell(q_0)}{\ell_{POV}(q_0)} &= \frac{\frac{k}{2}q_0^2 + \psi q_0 + \frac{(1+\phi)^2}{1+3\phi}\eta^{\frac{1}{1+\phi}}\left(\frac{\gamma\sigma^2}{2\phi V}\right)^{\frac{\phi}{1+\phi}} q_0^{\frac{1+3\phi}{1+\phi}}}{\frac{k}{2}q_0^2 + \psi q_0 + (1+\phi)\eta^{\frac{1}{1+\phi}}\left(\frac{\gamma\sigma^2}{6\phi V}\right)^{\frac{\phi}{1+\phi}} q_0^{\frac{1+3\phi}{1+\phi}}} \\[2mm]
&\geq \frac{\frac{(1+\phi)^2}{1+3\phi}\eta^{\frac{1}{1+\phi}}\left(\frac{\gamma\sigma^2}{2\phi V}\right)^{\frac{\phi}{1+\phi}} q_0^{\frac{1+3\phi}{1+\phi}}}{(1+\phi)\eta^{\frac{1}{1+\phi}}\left(\frac{\gamma\sigma^2}{6\phi V}\right)^{\frac{\phi}{1+\phi}} q_0^{\frac{1+3\phi}{1+\phi}}} \\[2mm]
&\geq \frac{1+\phi}{1+3\phi}3^{\frac{\phi}{1+\phi}}.
\end{aligned}
$$

Let us define $\Xi : \phi > 0 \mapsto \frac{1+\phi}{1+3\phi}3^{\frac{\phi}{1+\phi}}$. *We have*

$$
\Xi'(\phi) = 3^{\frac{\phi}{1+\phi}}\left(-\frac{2}{(1+3\phi)^2} + \frac{\log(3)}{(1+3\phi)(1+\phi)}\right).
$$

By solving the equation $\Xi'(\phi) = 0$, *we see that the unique extremum (minimum) of* Ξ *is* $\phi^* = \frac{2-\log(3)}{3\log(3)-2}$. *Therefore,*

$$
\frac{\ell(q_0)}{\ell_{POV}(q_0)} \geq \Xi(\phi^*) = \frac{e\log(3)}{2\sqrt{3}}.
$$

The above result highlights the fact that the use of a POV execution strategy should be similar to the use of an IS execution strategy when it comes to pricing a block trade.

8.5   Guaranteed VWAP contracts

So far, we have only dealt with the case of risk trades where the payment is made upfront. Other forms of risk trades do exist. An important case is that of Guaranteed VWAP contracts. In a Guaranteed VWAP contract, the intermediary does not pay the agent at time 0, but waits until a specified time T and pays a price for the portfolio that corresponds to the VWAP over the period $[0, T]$, plus or minus a premium decided upon at time 0.

The agent has the guarantee to obtain the VWAP (but pays for that, hence the premium). He does not bear the risk, contrary to what happens in the case of an agency trade with VWAP benchmark. Instead, the financial intermediary bears the risk and he has to pay the VWAP at the end of the period. In this section, we consider the problem faced by the intermediary in the case of a long single-stock portfolio with q_0 shares.

If we denote by π the premium received by the financial intermediary for the service, then his expected utility at time 0 is

$$\sup_{(v_t)_t \in \mathcal{A}} \mathbb{E}\left[-\exp(-\gamma(X_T - q_0(\mathrm{VWAP}_T - \pi)))\right], \qquad (8.9)$$

where

$$
\begin{aligned}
dq_t &= v_t dt, \\
dS_t &= \sigma dW_t + k v_t dt, \\
dQ_t &= V_t dt, \\
dX_t &= -v_t S_t dt - V_t L\left(\frac{v_t}{V_t}\right) dt,
\end{aligned}
$$

$$\mathcal{A} = \left\{ (v_t)_{t\in[0,T]} \in \mathbb{H}^0(\mathbb{R}, (\mathcal{F}_t)_t), \int_0^T v_t dt = -q_0, \int_0^T |v_t| dt \in L^\infty(\Omega) \right\},$$

and, as in Chapter 4,

$$\mathrm{VWAP}_T = \frac{\int_0^T S_t V_t dt}{\int_0^T V_t dt} = \frac{\int_0^T S_t dQ_t}{Q_T}.$$

If the intermediary was not participating in the Guaranteed VWAP contract, his utility would be

$$-\exp(-\gamma X_0). \qquad (8.10)$$

The premium $\pi = \pi_T(q_0)$ that equalizes Eq. (8.9) and Eq. (8.10) is given by

$$\pi_T(q_0) = \inf_{(v_t)_t \in \mathcal{A}} \frac{1}{\gamma q_0} \log\left(\mathbb{E}\left[\exp(-\gamma(X_T - X_0 - q_0 \mathrm{VWAP}_T))\right]\right).$$

To compute this premium, we just need to go back to Chapter 4 (Section 4.4.1) to see that

$$\pi_T(q_0) = \frac{k}{2}q_0 + \frac{1}{q_0}\min_{q \in \mathcal{C}_T} J_{\mathrm{VWAP}}(q),$$

where

$$J_{\mathrm{VWAP}}(q) =$$

$$\int_0^T \left(V_t L\left(\frac{q'(t)}{V_t}\right) - kq_0 \frac{V_t}{Q_T}(q_0 - q(t)) + \frac{1}{2}\gamma\sigma^2\left(q(t) - q_0\left(1 - \frac{Q_t}{Q_T}\right)\right)^2 \right) dt.$$

Theorem 4.3 states that if $k = 0$, then the minimizer in the above expression is the strategy $q^*(t) = q_0\left(1 - \frac{Q_t}{Q_T}\right)$, and we get

$$\pi_T(q_0) = Q_T L\left(-\frac{q_0}{Q_T}\right).$$

In particular, the premium is positive. In other words, the client has to pay for the service. This is however not true in general when $k > 0$. Guéant and Royer have shown in [91] that the premium can be negative when permanent market impact is important in magnitude. This happens because the intermediary can push down the price by selling rapidly at the beginning, and therefore push down the VWAP that he eventually needs to pay. This strategy is obviously risky and practitioners face a trade-off between sticking to the theoretically risk-free strategy $q(t) = q_0 \left(1 - \frac{Q_t}{Q_T}\right)$, and overselling at the beginning to push down the price and hope that it will stay quite stable afterwards.

Although the model we propose for Guaranteed VWAP contracts has a lot of drawbacks, it clearly highlights one of the reasons why buyers of Guaranteed VWAP contracts may sometimes be fooled by financial intermediaries. They indeed get the VWAP but this VWAP is influenced by the buying or selling pressure (as expected), and also by the way the execution is carried out, not in their favor.

8.6   Conclusion

The literature on optimal execution mainly focuses on strategies and tactics. In this chapter, we have seen how the Almgren-Chriss framework could be used to give a price to a block of shares or to price other types of risk trades, such as Guaranteed VWAP contracts. Pricing here is not based on replication, as in most of the literature. Instead, we rely on the theory of indifference pricing. In the following chapters, we will show how the Almgren-Chriss framework for execution and the indifference pricing approach can be combined to study derivative products. In Chapter 9, we focus on vanilla options, and we see that using execution models outside of the cash trading world enriches classical models of option pricing. It permits to make a difference between vanilla options with physical settlement and cash settlement, to discuss optimal hedging strategies in the presence of execution and transaction costs, etc. In Chapter 10, we also use the same tools to deal with complex share buy-back contracts.

The concept of risk-liquidity premium is very important when one wants to go beyond MtM valuation. It should be used whenever large positions are considered in financial models, be it to deal with derivative pricing/hedging, asset management,[14] or even the pricing of collateralized loans for instance. Although most quantitative analysts sweep this evidence under the rug, as they want to blindly apply the same linear models as others, liquidity has a value (and a price). The concept of risk-liquidity premium should remind quants that liquidity matters, and that Quantitative Finance has no reason to be full of linear equations; because reality is certainly nonlinear.

---

[14]In this book, we do not focus on asset management models, but we believe that in a next future, new asset management models, based on the ideas presented here, will be developed to penalize illiquidity.

Chapter 9

Option pricing and hedging with execution costs and market impact

You cannot swim for new horizons until you have courage to lose sight of the shore.

— William Faulkner

The academic literature on optimal execution addresses questions coming mainly from the cash-equity world: "how should one schedule the execution of a large order?", "how should one buy or sell stocks in practice?", "what should be the price of a block of shares?", etc. In this chapter, we show how the models developed for optimally executing orders can be used to price and hedge equity derivatives. More specifically, we show how the use of the Almgren-Chriss framework makes it possible to generalize classical option pricing models through the introduction of execution costs and permanent market impact, and by making a difference between options with physical settlement and options with cash settlement.

9.1 Introduction

9.1.1 Nonlinearity in option pricing

Is there still anything new that can be said about equity derivatives from a mathematical point of view? In recent years, quantitative analysts have not seen many new models implemented in the industry for pricing and hedging equity derivatives. Apart from local stochastic volatility (LSV) models that are now used by some banks for the pricing and hedging of equity derivatives (and for foreign exchange derivatives) – see for instance [94] – most of the tools used in practice, by most of the investment banks, are the same as five years ago, or even before the crisis. On the academic side however, many new models have been proposed, and many of them involve nonlinear equations.

LSV models are based on an old idea of Lipton – see his 2002 paper [139] – that makes it possible, in the same model, to benefit from both the perfect fit of a snapshot of the volatility surface (as with local volatility models), and a good modeling of the volatility surface dynamics (as with some stochastic volatility models). Although the basic ideas behind these models are not new, they have only started to be used very recently, following the works of Guyon and Henry-Labordère (see [94, 103]). These two authors have indeed showed how to proceed numerically to use them in practice, for any value of the spot-vol correlation – see also the works of Abergel and Tachet des Combes [2, 166]. The path-dependent volatility (PDV) approach of Guyon [93] is another example, far less implemented in practice, of a recent nonlinear approach that has the nice properties of both local and stochastic volatility models, without their main drawbacks. Both types of models are nonlinear: the price dynamics is not a standard stochastic differential equation (SDE), but instead a McKean SDE.

These recent improvements in the modeling of option prices are based on nonlinear equations, whereas common approaches are linear.[1] One may argue that nonlinear models are not new in finance. The most common nonlinear models are those dealing with optimal stopping problems, in particular for the early exercise of Bermudan or American options. However, apart from these models, most of the nonlinear models in finance are seen as curiosities rather than as practical tools that could be used in daily business. This is true of the old uncertainty volatility model of Avellaneda et al. [11], which leads to the nonlinear Black-Scholes-Barenblatt partial differential equation. This is also true of many super-replication models. This is eventually the case of most of the models with transaction costs and those focused on the feedback effect – see the book [95] for a review of many nonlinear problems in finance.

There are many reasons why truly nonlinear models are not often used by practitioners. The main reason is that options are very seldom managed on a stand-alone basis, but instead within a book. In other words, for a nonlinear option pricing model to be really efficient, it should not price a given option in isolation, but instead price that option given the current state of the book. With linear models, such a problem does not exist, by definition. For managing a portfolio of derivatives, practitioners clearly find it easier to sum up Greeks – coming from linear equations – than considering complex nonlinear equations. LSV models are nonlinear models, but, because the backward pricing PDE in these models is linear, traders can use them and continue to reason in terms of Greeks. Another reason why nonlinear models are seldom used is because they are harder to solve numerically than their linear rivals.[2] Another important reason why nonlinear models are rarely used is because almost no-

---

[1]LSV models are nonlinear models, but the backward pricing partial differential equation (PDE) is a linear one – contrary to the forward PDE, which is not.

[2]The models for American options are an exception. Equations are nonlinear, because of the free boundary, but there is no difficulty as far as numerics is concerned.

body in business has an incentive to use new models – see for instance the thought-provoking book of Wilmott [173]. When a new model appears, it usually takes several years before anybody implements it in practice. A plausible explanation is that models need to be tested very carefully. However, most of the models used in practice are not tested – at least not in the same way cars and aeroplanes are tested by engineers.[3] In fact, the current situation is very close to a suboptimal Nash equilibrium in terms of modeling choice: nobody wants to be the first one using a new model, by fear of being very wrong alone.

9.1.2 Liquidity sometimes matters

An important consequence of the above discussion is that the option pricing models which take into account liquidity issues are very marginally used, because almost are all nonlinear. However, the baby should not be thrown out with the bathwater.

Liquidity is certainly not the most important factor for successfully managing a book of vanilla options, but it is important to manage options with large nominal, especially if they are managed on a stand-alone basis, as it is sometimes the case in corporate deals.

Liquidity issues often strike back, sometimes dramatically, near the expiration date of options, when the price oscillates around the strike. A recent example is given in Figures 9.1 and 9.2. On July 19, 2012, the price of several American blue-chip stocks, including Coca-Cola and Apple, but also McDonald's and IBM, exhibited a surprising saw-tooth pattern: the stock price reached a local minimum every hour, alternatively with a local maximum after the first half-hour of each hour. Lehalle et al. analyzed in detail that specific day in [129], and concluded that the patterns observed on that day (one day before the third Friday of the month of July) were certainly due to a trader who blindly Δ-hedged a large position in options. In the present case, the trader was certainly replicating a negative Γ position, and was making a trading decision every half-hour. The most plausible story is the following. Step 1: at 10:00, his Δ-hedging strategy consisted in buying shares and he certainly used an algorithm (for instance a VWAP algorithm) to execute his orders over 30 minutes. At 10:30, the price had increased because his buying order had pushed up the price. Step 2: At 10:30, because he was replicating a negative Γ position (or having a long Γ position), the price increase led to an increase in his Δ. This forced him to sell shares to remain Δ-hedged. Therefore, he sent a sell order to be executed over the next half-hour. Because of market impact, the stock price went down until 11:00. Step 3 (similar to Step 1, but one hour

---

[3]Should they be tested, it would turn out that they do not work at the level of single options. Nevertheless, most of the time, models do the job at the level of books.

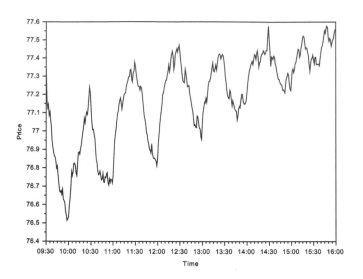

FIGURE 9.1: Saw-tooth patterns on American blue-chip stocks. The case of Coca-Cola on July 19, 2012.

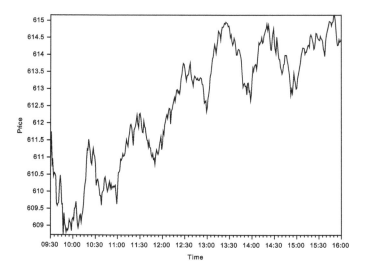

FIGURE 9.2: Saw-tooth patterns on American blue-chip stocks. The case of Apple on July 19, 2012.

later): the trader had to buy shares to Δ-hedge his position (because the price had moved down), pushing up the price and forcing him, half an hour later, to sell part of what he had bought, in order to apply his nonsense Δ-hedging strategy. And so on.

This example shows that liquidity is not only for textbooks: it should always be taken into account in the case of options with a huge nominal. Although they are rarely used in practice, numerous models have been proposed to include transaction costs and market impact. One of the oldest and most famous models to account for transaction costs is the model proposed by Leland [130], which dates back to 1985. Other old models with fixed transaction costs or transaction costs proportional to the traded volume include (among many others) the model of Barles and Soner [18], or the model proposed by Cvitanić, Pham, and Touzi [57]. Another strand of literature dealing with liquidity is associated with what is usually called the supply curve approach. Çetin, Jarrow, and Protter [46] introduced this approach, in which traders are not price takers, but instead trade at a price that depends on the volume they trade. In its basic form, this approach leads to the same prices as in the Black-Scholes model, but the model was improved later on. For instance, Çetin, Soner, and Touzi in [47] got interesting results through the addition of Γ constraints.

More related to the saw-tooth pattern of July 2012, another old strand of literature has been dealing with the impact of Δ-hedging on the dynamics of the price of the underlying, and on its impact on the option price. This literature on the feedback effect includes for instance [152], [159], and [174].

The literature on optimal execution deals with execution costs and market impact. It is clear from the previous paragraphs that the ideas coming from this literature should be useful to deal with the costs related to hedging, and the effect of hedging on the price of the underlying. Several models based on the ideas of the Almgren-Chriss model have been proposed to deal with option pricing issues. Rogers and Singh [158] were certainly the first ones to propose a simple model with quadratic execution costs, but no permanent market impact. In their model, the optimal hedging strategy is approximated by a mean-reverting strategy around the classical Δ. Lions and Lasry [138] proposed a model with no execution costs but linear permanent market impact. Another recent approach with permanent market impact only is the one proposed by Bouchard et al. in [30]. Guéant and Pu proposed in [89] a model with both permanent market impact and execution costs, based on the Almgren-Chriss framework with a CARA utility function. Li and Almgren [134] proposed another model with both features, but with a different objective function.

The model we propose in this chapter is inspired from [89]. It is a pure application of the Almgren-Chriss framework of Chapter 3. In other words, it (almost) does not use any idea from classical option pricing. We aim at (i) finding the optimal hedging strategy for a vanilla option, (ii) pricing vanilla options in that model by using the indifference pricing approach, and (iii) showing that we can easily distinguish in our model the case of options with physical delivery from the case of options with cash settlement – two cases that are never considered separately in the classical literature, because one evaluates stocks at their MtM price.

9.2 The model in continuous time

9.2.1 Setup of the model

We consider a trader with a position (long or short) in a European vanilla option, that is, a call or a put on a stock. We consider that the trader wants to hedge his position. For that purpose, he can buy and sell stocks between time 0 and time T (the expiration date of the option).

To model the price process of the underlying and the hedging process, we use the Almgren-Chriss framework.

The number of shares in the hedging portfolio is modeled by a process $(q_t)_{t \in [0,T]}$. Its dynamics is

$$dq_t = v_t dt, \qquad (9.1)$$

where $(v_t)_{t \in [0,T]}$ is a control in the set $\mathcal{A}_{\rho_{\max}}$ of admissible controls defined by

$$\mathcal{A}_{\rho_{\max}} = \left\{ (v_t)_{t \in [0,T]} \in \mathbb{H}^0(\mathbb{R}, (\mathcal{F}_t)_t), |v_t| \leq \rho_{\max} V_t, \text{a.e.} \right\},$$

where $\rho_{\max} > 0$ is an upper bound for the participation rate, and $(V_t)_{t \in [0,T]}$ is the market volume process, which represents the volume traded by other agents. As in the previous chapters, we assume that $(V_t)_{t \in [0,T]}$ is a deterministic nonnegative[4] and bounded process. However, unlike what happens in execution models, we only assume that $(V_t)_{t \in [0,T]}$ is continuous by parts, because the time interval $[0,T]$ spans several days (and nights, where the market volume is 0).

The mid-price of the stock is modeled with the process $(S_t)_{t \in [0,T]}$ following the dynamics

$$dS_t = \sigma dW_t + k v_t dt. \qquad (9.2)$$

---

[4]Assuming that the market volume process is nonnegative, instead of positive, does not raise any problem. Furthermore, the night can then be modeled.

As in the Almgren-Chriss execution model, we consider that there is some form of permanent impact on the stock price. Here, it is the hedging strategy that impacts the price of the underlying, and this impact must be taken into account in the hedging strategy (for instance to avoid the saw-tooth pattern exhibited in Figures 9.1 and 9.2).[5]

As in the Almgren-Chriss execution model of Chapter 3, we also consider that the stock price, in the absence of market impact, is normally distributed. In particular, the stock price does not follow the classical log-normal dynamics of the Black and Scholes option pricing model. This may be a problem for long-maturity options, but not a real problem for short-maturity options. Anyhow, using a normal model instead of the classical log-normal one, introduces a form of negative skew, which is not bad when it comes to option pricing.[6]

As far as execution costs are concerned, we use an execution cost function L with the same assumptions as in Chapter 3. Therefore, the cash account process $(X_t)_{t \in [0,T]}$ evolves as

$$dX_t = -v_t S_t dt - V_t L \left(\frac{v_t}{V_t} \right) dt. \tag{9.3}$$

Now, we consider what happens at time T, when the option expires, and we have to distinguish between the different types of options.

Let us first detail the case of a trader who has sold a call option with strike K and nominal Q.

If the stock price S_T is below the strike K, then the option expires worthless and we assume that the trader liquidates his portfolio (equivalently, we can assume that we evaluate the portfolio using a block trade price – see Chapter 8). In other words, his terminal wealth is

$$X_T + q_T S_T - \ell(q_T),$$

where $\ell(q_T)$ is a risk-liquidity premium to unwind a position with q_T shares (see Chapter 8).

If the stock price S_T is above the strike K, then the terminal wealth depends on the type of settlement.

In the case of a physical settlement (PS), the counterpart pays K for each of the Q shares he receives. The trader may not have Q shares in his portfolio

---

[5]Permanent market impact raises questions about manipulation because the trader can, theoretically, push the price up or down and impact the outcome of the option contract.

[6]The differences between the log-normal model of Black and Scholes and the normal model of Bachelier are discussed in [160].

at time T (he has q_T shares), but we assume that he can buy instantaneously the $Q - q_T$ shares he does not have for the amount $(Q - q_T)S_T + \ell(Q - q_T)$. Therefore, the final wealth in the case of physical settlement is

$$X_T + QK - (Q - q_T)S_T - \ell(Q - q_T).$$

In the case of cash settlement (CS), the trader pays $Q(S_T - K)$ to the counterpart who has bought the call option, and we assume that he liquidates his portfolio (or at least we replace the MtM price of the portfolio by its block trade price, as in Chapter 8). Therefore, the terminal wealth is

$$X_T + q_T S_T - \ell(q_T) - Q(S_T - K).$$

Wrapping up, we get that the terminal wealth of the trader is

$$X_T + q_T S_T - Q(S_T - K)_+ - \ell(q_T)1_{\text{CS}} - (\ell(Q - q_T)1_{S_T > K} + \ell(q_T)1_{S_T \leq K})\,1_{\text{PS}},$$

in the case of a short position in a call option with strike K, maturity T, and nominal Q.

We see in particular that the strike K and the maturity T are not the only two relevant parameters for pricing and hedging a vanilla option: the nominal Q also matters, unlike what happens with a linear model. We also see that, in the case of physical settlement, the terminal wealth is not continuous at $S_T = K$.

The other types of vanilla options can be considered in the same way. Table 9.1 exhibits the different terminal wealths corresponding to the different options and types of settlement. More complex options can be considered very easily, using the same reasoning.

In all cases, the terminal wealth is of the form

$$X_T + q_T S_T + \Pi(q_T, S_T),$$

for some function Π, that depends on the type of option, the position, and the type of settlement.

To find the optimal hedging strategy, we consider the same expected utility criterion as in Chapter 3. In other words, we consider the following problem:

$$\sup_{(v_t)_t \in \mathcal{A}_{\rho\max}} \mathbb{E}\left[-\exp(-\gamma(X_T + q_T S_T + \Pi(q_T, S_T)))\right].$$

As far as pricing is concerned, we focus on the indifference price (for the trader) of a contract where (i) the trader writes the option contract and the counterpart pays a price P, and (ii) the counterpart gives q_0 shares to the trader and receives $q_0 S_0$ in cash from the trader.

TABLE 9.1: Terminal wealth for different types of options.

Option	Position	Settlement	Terminal wealth
Call	Short	PS	$X_T + q_T S_T - Q(S_T - K)_+$ $- (\ell(Q - q_T)1_{S_T>K} + \ell(q_T)1_{S_T\leq K})$
		CS	$X_T + q_T S_T - Q(S_T - K)_+$ $-\ell(q_T)$
	Long	PS	$X_T + q_T S_T + Q(S_T - K)_+$ $- (\ell(Q + q_T)1_{S_T>K} + \ell(q_T)1_{S_T\leq K})$
		CS	$X_T + q_T S_T + Q(S_T - K)_+$ $-\ell(q_T)$
Put	Short	PS	$X_T + q_T S_T - Q(S_T - K)_-$ $- (\ell(Q + q_T)1_{S_T<K} + \ell(q_T)1_{S_T\geq K})$
		CS	$X_T + q_T S_T - Q(S_T - K)_-$ $-\ell(q_T)$
	Long	PS	$X_T + q_T S_T + Q(S_T - K)_-$ $- (\ell(Q - q_T)1_{S_T<K} + \ell(q_T)1_{S_T\geq K})$
		CS	$X_T + q_T S_T + Q(S_T - K)_-$ $-\ell(q_T)$

We consider such a contract, instead of the option contract itself, because of the initial Δ (in the wording of classical models). If the nominal is large, then building quickly the initial Δ-hedging portfolio (of the classical approach) may be impossible. In practice, for some corporate deals, the counterpart indeed provides the bank with stocks.

The indifference price P (positive or negative[7]) is the price that makes

$$\sup_{(v_t)_t \in \mathcal{A}_{\rho\max}} \mathbb{E}\left[-\exp(-\gamma(X_T + q_T S_T + \Pi(q_T, S_T) + P - q_0 S_0))\right]$$

equal to the utility in the no-deal case, i.e.,

$$-\exp(-\gamma X_0).$$

In other words, the price P is given by

$$P = \inf_{(v_t)_t \in \mathcal{A}_{\rho\max}} \frac{1}{\gamma} \log \mathbb{E}\left[\exp(-\gamma(X_T - X_0 + q_T S_T - q_0 S_0 + \Pi(q_T, S_T)))\right].$$

(9.4)

The pricing and hedging problems are related but, unlike in the Black-Scholes model or in other classical models, the price is not the cost of replicating (hedging) the option. In classical models, the price is simply the cost

---

[7]With our convention, the price of a short (long) position should be positive (negative). It corresponds to the minimum amount of cash the trader needs to receive at time 0 to accept the deal.

of production of the option (the option being produced at no risk because of the complete market assumption). Here, the market is incomplete because of execution costs, and therefore the price of the contract takes account of both the cost for "producing" the option and the risk of the production process.

9.2.2 Towards a new nonlinear PDE for pricing

9.2.2.1 The Hamilton-Jacobi-Bellman equation

To solve both pricing and hedging problems, we introduce the value function

$$u(t, x, q, S) = \sup_{v \in \mathcal{A}_t} \mathbb{E}\left[-\exp\left(-\gamma\left(X_T^{t,x,v} + q_T^{t,q,v} S_T^{t,S,v} + \Pi(q_T^{t,q,v}, S_T^{t,S,v})\right)\right)\right],$$

where

$$\mathcal{A}_t := \left\{(v_s)_{s \in [t,T]} \in \mathbb{H}^0(\mathbb{R}, (\mathcal{F}_s)_{t \le s \le T}), |v_s| \le \rho_{\max} V_s, \text{a.e. in } (t, T) \times \Omega\right\},$$

and

$$dX_s^{t,x,v} = -v_s S_s^{t,S,v} ds - V_s L\left(\frac{v_s}{V_s}\right) ds, \quad X_t^{t,x,v} = x,$$

$$dq_s^{t,q,v} = v_s ds, \quad q_t^{t,q,v} = q,$$

$$dS_s^{t,S,v} = \sigma dW_s + k v_s ds, \quad S_t^{t,S,v} = S.$$

The Hamilton-Jacobi-Bellman (HJB) equation associated with this problem is

$$-\partial_t u - \frac{1}{2}\sigma^2 \partial_{SS}^2 u$$

$$- \sup_{|v| \le \rho_{\max} V_t} \left\{v \partial_q u + \left(-vS - L\left(\frac{v}{V_t}\right) V_t\right) \partial_x u + k v \partial_S u\right\} = 0, \qquad (9.5)$$

with the terminal condition

$$u(T, x, q, S) = -\exp\left(-\gamma\left(x + qS + \Pi(q, S)\right)\right).$$

9.2.2.2 The pricing PDE

To get the pricing PDE, the first step of the reasoning consists in computing

$$X_T^{t,x,v} + q_T^{t,q,v} S_T^{t,S,v} + \Pi(q_T^{t,q,v}, S_T^{t,S,v})$$

as a function of x, q, and S, by using Eqs. (9.1), (9.2), and (9.3).

We have

$$
X_T^{t,x,v} + q_T^{t,q,v} S_T^{t,S,v} + \Pi(q_T^{t,q,v}, S_T^{t,S,v})
$$

$$
= x + \int_t^T \left(-v_s S_s^{t,S,v} - V_s L\left(\frac{v_s}{V_s}\right) \right) ds
$$

$$
+ qS + \int_t^T v_s S_s^{t,S,v} ds + \int_t^T \sigma q_s^{t,q,v} dW_s
$$

$$
+ k \int_t^T v_s q_s^{t,q,v} ds + \Pi(q_T^{t,q,v}, S_T^{t,S,v})
$$

$$
= x + qS - \int_t^T V_s L\left(\frac{v_s}{V_s}\right) ds + \int_t^T \sigma q_s^{t,q,v} dW_s
$$

$$
+ \frac{k}{2} q_T^{t,q,v\,2} - \frac{k}{2} q^2 + \Pi(q_T^{t,q,v}, S_T^{t,S,v}). \tag{9.6}
$$

Using the definition of u, we have from Eq. (9.6) that

$$
u(t, x, q, S) = -\exp\left(-\gamma(x + qS - \theta(t, q, S))\right), \tag{9.7}
$$

where

$$
\theta(t, q, S) = \inf_{v \in \mathcal{A}_t} \frac{1}{\gamma} \log \left(\mathbb{E}\left[\exp\left(-\gamma \left(-\int_t^T V_s L\left(\frac{v_s}{V_s}\right) ds \right. \right. \right. \right.
$$

$$
\left. \left. \left. \left. + \int_t^T \sigma q_s^{t,q,v} dW_s + \frac{k}{2} q_T^{t,q,v\,2} - \frac{k}{2} q^2 + \Pi(q_T^{t,q,v}, S_T^{t,S,v}) \right) \right) \right] \right).
$$

From Eqs. (9.4) and (9.7), we know that $\theta(0, q_0, S_0)$ is the price of the option contract described previously. In particular, this price depends on q_0. This is natural because, in practice, it is easier to hedge an option if one starts with the "correct" initial Δ.

Now, we want to characterize θ by using a partial differential equation. For this purpose, we need to rely on the theory of viscosity solution. We know that u is a viscosity solution of the HJB equation (9.5). Therefore, by using the change of variables of Eq. (9.7), we get that θ is a viscosity solution[8] of the PDE (9.8):

$$
-\partial_t \theta - \frac{1}{2} \sigma^2 \partial_{SS}^2 \theta - \frac{1}{2} \gamma \sigma^2 (\partial_S \theta - q)^2
$$

$$
- \inf_{|v| \le \rho_{\max} V_t} \left\{ v \partial_q \theta + L\left(\frac{v}{V_t}\right) V_t - kv(q - \partial_S \theta) \right\} = 0, \tag{9.8}
$$

with $\theta(T, q, S) = -\Pi(q, S)$.

---

[8] Here, we do not prove a uniqueness result. In fact, some research still needs to be carried out in order to have a complete and rigorous assessment.

In the absence of permanent impact, Eq. (9.8) gets simplified into

$$-\partial_t \theta - \frac{1}{2}\sigma^2 \partial_{SS}^2 \theta - \frac{1}{2}\gamma\sigma^2 (\partial_S \theta - q)^2 + V_t H_{\rho_{\max}}(-\partial_q \theta) = 0, \qquad (9.9)$$

where

$$H_{\rho_{\max}}(p) = \sup_{|\rho| \le \rho_{\max}} \rho p - L(\rho).$$

In fact, up to a change of variables, this equation is still correct, even when permanent market impact is taken into account.

If indeed we introduce

$$\widehat{\theta} : \left(t, q, \widehat{S}\right) \mapsto \frac{k}{2}(q_0^2 - q^2) + \theta\left(t, q, \widehat{S} + k(q - q_0)\right),$$

then $\widehat{\theta}$ is viscosity solution of the same equation as above, that is,

$$-\partial_t \widehat{\theta} - \frac{1}{2}\sigma^2 \partial_{\widehat{S}\widehat{S}}^2 \widehat{\theta} - \frac{1}{2}\gamma\sigma^2 (\partial_{\widehat{S}} \widehat{\theta} - q)^2 + V_t H_{\rho_{\max}}(-\partial_q \widehat{\theta}) = 0,$$

but with terminal condition

$$\widehat{\theta}(T, q, \widehat{S}) = \frac{k}{2}(q_0^2 - q^2) - \Pi(q, \widehat{S} + k(q - q_0)).$$

Furthermore, the indifference price P of the contract is $P = \theta(0, q_0, S_0) = \widehat{\theta}(0, q_0, S_0)$. Therefore, the influence of the permanent market impact can be analyzed as a modification of the terminal condition.

It is noteworthy that the new variable $\widehat{S} = S - k(q - q_0)$ is the price from which we have removed the permanent market impact.

9.2.3   Comments on the model and the pricing PDE

We now analyze the different terms of the PDE (9.9):

- The term $-\partial_t \theta - \frac{1}{2}\sigma^2 \partial_{SS}^2 \theta$ corresponds to the backward heat equation associated with the dynamics of the stock price. In particular, the PDE (9.9) generalizes the heat equation that one gets with the classical Bachelier model.

- The term $V_t H_{\rho_{\max}}(-\partial_q \theta)$ comes from execution costs. It is the same as in the Hamilton-Jacobi equation associated with the optimal liquidation (with an IS strategy) – see Eq. (8.6). This term is important for at least two reasons. First, it quantifies illiquidity, that is, the difficulty of buying or selling shares. Second, it gives the optimal hedging strategy

$$v^*(t, q_t, S_t) = V_t H'_{\rho_{\max}}(-\partial_q \theta(t, q_t, S_t)),$$

or

$$v^*(t, q_t, S_t) = V_t H'_{\rho_{\max}}(-\partial_q \widehat{\theta}(t, q_t, S_t - k(q_t - q_0))),$$

in the general case with permanent market impact.

- The fact that the portfolio of shares is used for hedging is related to the term $\frac{1}{2}\gamma\sigma^2(\partial_S\theta - q)^2$. With this term, we see that the risk is related to the difference between the number of shares q in the portfolio, and the sensitivity of θ (remember θ is intimately linked to the price of the option) to the price S of the underlying. Although there is no Δ in our model, this term is a form of measurement of the mis-hedge.

Eq. (9.9) is evidently a nonlinear PDE. Furthermore, it is important to highlight that, compared to most classical pricing PDEs (even nonlinear ones), there are not only two variables (time and price), but an additional one: q. In practice, this means that numerical schemes will be more complicated and more time-consuming.

Another important point regarding the PDE (9.9) is linked to the nominal Q of the option. Because Eq. (9.9) is nonlinear, the price of an option with nominal Q is not Q times the price of the same option with nominal 1. It is however interesting to factor out the nominal, and to consider the normalized function

$$\widehat{\theta}_Q : (t, \widehat{q}, \widehat{S}) \mapsto \frac{k}{2}Q(\widehat{q}_0^2 - \widehat{q}^2) + \frac{1}{Q}\theta\left(t, Q\widehat{q}, \widehat{S} + kQ(\widehat{q} - \widehat{q}_0)\right),$$

where $\widehat{q}_0 = \frac{q_0}{Q}$ is the normalized initial condition for the hedging portfolio.

It is straightforward to verify that $\widehat{\theta}_Q$ is viscosity solution of

$$-\partial_t\widehat{\theta}_Q - \frac{1}{2}\sigma^2\partial_{\widehat{S}\widehat{S}}^2\widehat{\theta}_Q - \frac{1}{2}\gamma\sigma^2 Q(\partial_{\widehat{S}}\widehat{\theta}_Q - \widehat{q})^2 + \frac{V_t}{Q}H_{\rho_{\max}}(-\partial_{\widehat{q}}\widehat{\theta}_Q) = 0, \quad (9.10)$$

with terminal condition

$$\widehat{\theta}_Q(T, \widehat{q}, \widehat{S}) = \frac{k}{2}Q(\widehat{q}_0^2 - \widehat{q}^2) - \frac{1}{Q}\Pi(N\widehat{q}, \widehat{S} + kQ(\widehat{q} - \widehat{q}_0)).$$

Furthermore, the optimal hedging strategy is

$$v^*(t, q_t, S_t) = V_t H'_{\rho_{\max}}(-\partial_{\widehat{q}}\widehat{\theta}(t, \frac{q_t}{Q}, S_t - k(q_t - q_0))).$$

Changing the nominal boils down therefore to rescaling the risk aversion parameter γ, the market volume process $(V_t)_t$, the permanent market impact parameter k, and, obviously, the function Π, which incorporates the payoff (in the classical sense), and the risk-liquidity premia. Obviously, one reasons then on a rescaled hedging portfolio, hence the variable $\hat{q} = \frac{q}{Q}$.

The PDEs (9.8), (9.9), and (9.10) can be solved numerically using finite difference schemes for nonlinear equations (see [89]). Interestingly, we can also build a discrete-time version of the model and propose a tree-based method to solve the nonlinear problem in a very simple way.

Before moving to the discrete-time model, we need to highlight that the model can be generalized to include interest rates and a drift in the dynamics of the stock price. Unlike what happens in a complete market model, the price of the option and the hedging strategy are impacted by the presence of a drift. Because we cannot exactly replicate the option, the probabilities that an option expires either in the money or worthless do matter. Therefore, the drift plays a role. We refer the reader to [89] for more details on the roles of the drift and the interest rate in this model.

9.3   The model in discrete time

In the previous section, we have considered a continuous-time model. If we solve numerically one of the PDEs (9.8), (9.9), or (9.10), then we can compute the optimal hedging strategy. Yet, in practice, we are going to send orders to buy and sell shares at discrete times. In this section, we directly consider that we are making decisions at discrete times, and we build a discrete-time counterpart of the above model, where hedging decisions occur at fixed time intervals. In practice, it may not be optimal to rehedge at fixed time intervals, but we also see our discrete-time model as a good approximation of the continuous-time one, especially when it comes to building numerical methods.

9.3.1   Setup of the model

To build a discrete-time model, we divide the time window $[0, T]$ into N slices of length Δt. We denote by $t_0 = 0 < \ldots < t_n = n\Delta t < \ldots < t_N = N\Delta t = T$ the associated subdivision of the time interval $[0, T]$.

At the beginning of each time interval $[t_n, t_{n+1}]$, the trader chooses the number of shares he will buy (or sell) over this time interval. We denote by $v_n \Delta t$ the number of shares bought by the trader between t_n and t_{n+1} (if this number is negative, it corresponds to selling shares instead of buying).

The dynamics for the number of shares in the trader's portfolio is therefore

$$q_{n+1} = q_n + v_n \Delta t, \quad 0 \le n < N.$$

The price of the stock has the following dynamics

$$S_{n+1} = S_n + \sigma \sqrt{\Delta t} \epsilon_{n+1}, \quad 0 \le n < N,$$

where $\epsilon_1, \ldots, \epsilon_N$ are i.i.d. random variables with mean 0 and variance 1.

The dynamics of the cash account is similar to Eq. (9.3):

$$X_{n+1} = X_n - v_n S_n \Delta t - L\left(\frac{v_n}{V_{n+1}}\right) V_{n+1} \Delta t, \quad 0 \le n < N,$$

where $V_{n+1}\Delta t$ is the market volume over $[t_n, t_{n+1}]$.[9]

At least two remarks can be made on this model.

First, we have chosen not to include permanent market impact. We have seen in the continuous-time model that permanent market impact can be modeled by a change in the terminal condition. Therefore, to be parsimonious, we prefer to put aside the permanent market impact. Furthermore, in discrete time, if one directly considers the permanent component of market impact, then numerical methods are far more complex.

The second remark is that, as in Eq. (3.18), the cost paid to buy or sell shares is relative to S_n, which is the price at the beginning of the time interval $[t_n, t_{n+1}]$ over which the trades occur. In particular, we choose to neglect the price risk over each time slice.[10]

Using the same reasoning as in the previous section, we can prove that the wealth at time $t_N = T$, just after the option expiration, is of the form

$$X_N + q_N S_N + \Pi(q_N, S_N),$$

where Π depends on the type of option, the position, and the type of settlement (see Table 9.1).

Our goal is then to maximize

$$\mathbb{E}\left[-\exp\left(-\gamma\left(X_N + q_N S_N + \Pi(q_N, S_N)\right)\right)\right],$$

over

$$\mathcal{A}_d = \{(v_n)_{0 \le n < N}, |v_n| \le \rho_{\max} V_{n+1}, \forall n\}.$$

---

[9]The market volume process is assumed to be deterministic.

[10]This price risk can be included artificially in the execution function L, with a premium. Furthermore, it must be noted that the problem at stake is a problem over several days, weeks, or months. Subsequently, the function L should be adapted to the time interval Δt.

9.3.2   A new recursive pricing equation

The above problem can be solved using the classical tools of discrete-time optimization. We introduce a sequence of functions, which constitutes the discrete counterpart of the value function in continuous time:

$$\forall n \in \{0, \ldots, N\}, \qquad u^n(x, q, S) =$$

$$\mathbb{E}\left[-\exp\left(-\gamma \left(X_N + q_N S_N + \Pi(q_N, S_N)\right)\right) | X_n = x, q_n = q, S_n = S\right].$$

The functions $(u^n)_n$ satisfy the following Bellman equations:

$$\forall n \in \{0, \ldots, N-1\},$$

$$u^n(x, q, S) = \sup_{|v| \le \rho_{\max} V_{n+1}} \mathbb{E}\left[u^{n+1}\left(x - vS\Delta t - L\left(\frac{v}{V_{n+1}}\right) V_{n+1}\Delta t,\right.\right.$$

$$\left.\left. q + v\Delta t, S + \sigma\sqrt{\Delta t}\epsilon_{n+1}\right)\right], \tag{9.11}$$

with the terminal condition

$$u^N(x, q, S) = -\exp\left(-\gamma \left(x + qS + \Pi(q, S)\right)\right).$$

As in Eq. (9.7), we use a change of variables:

$$u^n(x, q, S) = -\exp\left(-\gamma \left(x + qS - \theta^n(q, S)\right)\right).$$

Using Eq. (9.11), we see that the sequence of functions $(\theta^n)_n$ satisfies the following equations:

$$\forall n \in \{0, \ldots, N-1\},$$

$$\theta^n(q, S) = \frac{1}{\gamma} \inf_{|v| \le \rho_{\max} V_{n+1}} \log \mathbb{E}\left[\exp\left(\gamma\left(L\left(\frac{v}{V_{n+1}}\right) V_{n+1}\Delta t\right.\right.\right.$$

$$\left.\left.\left. -(q + v\Delta t)\sigma\sqrt{\Delta t}\epsilon_{n+1} + \theta^{n+1}(q + v\Delta t, S + \sigma\sqrt{\Delta t}\epsilon_{n+1})\right)\right)\right], \tag{9.12}$$

with the terminal condition

$$\theta^N(q, S) = -\Pi(q, S).$$

As far as pricing is concerned, the price of the option contract (with the initial exchange of q_0 shares against the amount $q_0 S_0$ in cash) is given by $\theta^0(q_0, S_0)$. As far as hedging is concerned the optimal number of shares to buy (or sell) over $[t_n, t_{n+1}]$ is given by $v_n^*(q, S)\Delta t$, where $v_n^*(q, S)$ is a minimizer in Eq. (9.12), i.e.,

$$v_n^*(q, S) \in \operatorname{argmin}_{|v| \le \rho_{\max} V_{n+1}} \log \mathbb{E}\left[\exp\left(\gamma\left(L\left(\frac{v}{V_{n+1}}\right) V_{n+1}\Delta t\right.\right.\right.$$

$$\left.\left.\left. -(q + v\Delta t)\sigma\sqrt{\Delta t}\epsilon + \theta^{n+1}(q + v\Delta t, S + \sigma\sqrt{\Delta t}\epsilon)\right)\right)\right].$$

9.4   Numerical examples

To price and hedge an option with our model, we need to rely on numerical methods to solve one of the PDEs (9.8), (9.9), or (9.10), or the recursive equation (9.12). Here, we focus on the latter. By specifying in a specific way the distribution of $(\epsilon_1, \ldots, \epsilon_N)$, we obtain a very simple trinomial-tree method.

9.4.1   A trinomial tree

We now consider that the $(\epsilon_n)_n$ are i.i.d. with the following distribution:

$$
\epsilon_n = \begin{cases} \bar{\epsilon} & \text{with probability } \frac{1-p}{2}, \\ 0 & \text{with probability } p, \\ -\bar{\epsilon} & \text{with probability } \frac{1-p}{2}, \end{cases}
$$

where $\bar{\epsilon} = \frac{1}{\sqrt{1-p}}$, and where $p \in (0,1)$ can be chosen freely (for instance $p = \frac{2}{3}$ to fit the kurtosis of a normally distributed variable, or $p = \frac{1}{2}$, as in [89]).

In order to solve the recursive equation (9.12), we just need to compute $\theta^n(\cdot, S)$, for (n, S) corresponding to a trinomial tree – the first two levels are represented below.[11]

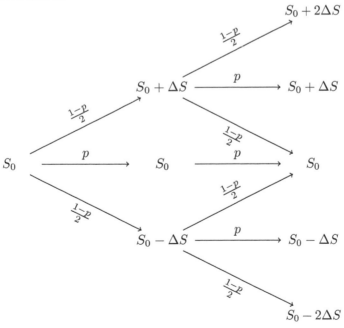

---

[11]$\Delta S = \sigma\sqrt{\Delta t}\bar{\epsilon}.$

Contrary to what happens with classical tree methods, we need to compute more than one value at each node. At each node (n, S), we need indeed to compute a function $q \mapsto \theta^n(q, S)$. In practice, we consider a discrete set of values of q, relevant with the problem. For instance, if one is short a call (or long a put) of nominal Q, then one has to choose a set of the form

$$\mathcal{Q} = \{0, \Delta q, 2\Delta q, \ldots, Q\},$$

where Q is a multiple of Δq. On the contrary, if one is long a call (or short a put) of nominal Q, then one has to choose a set of the form

$$\mathcal{Q} = \{0, -\Delta q, -2\Delta q, \ldots, -Q\}.$$

In practice, with the above approach, we approximate the values of $\theta^n(q, S)$ for (n, S) corresponding to the nodes of the trinomial tree and $q \in \mathcal{Q}$. This is done backward in time by using Eq. (9.12), that is,[12]

$$
\theta^n(q, S) = \inf_{\mathsf{q} \in \mathcal{Q}, |q - \mathsf{q}| \leq \rho_{\max} V_{n+1} \Delta t} \left\{ L\left(\frac{q - \mathsf{q}}{V_{n+1} \Delta t} \right) V_{n+1} \Delta t \right.
$$
$$
+ \frac{1}{\gamma} \log \left[\frac{1 - p}{2} \exp \left(\gamma \left(-\mathsf{q} \Delta S + \theta^{n+1}(\mathsf{q}, S + \Delta S) \right) \right) \right.
$$
$$
+ p \exp \left(\gamma \left(\theta^{n+1}(\mathsf{q}, S) \right) \right)
$$
$$
\left. \left. + \frac{1 - p}{2} \exp \left(\gamma \left(\mathsf{q} \Delta S + \theta^{n+1}(\mathsf{q}, S - \Delta S) \right) \right) \right] \right\},
$$

and the terminal condition

$$\theta^N(q, S) = -\Pi(q, S).$$

9.4.2 Hedging a call option with physical delivery

We now use the above trinomial-tree approach to compute the optimal hedging strategy associated with a short position in a call, in the case of physical delivery. We also compute the indifference price of the contract for the trader.

The parameters for the underlying stock are[13]

- $S_0 = 45$ €,

---

[12]There is an abuse of notation here, since we use the same notation both for the function and for its approximation.

[13]We assume that there is no permanent market impact in this example. See [89] for a discussion on permanent market impact.

- $\sigma = 0.6 \ €\cdot\text{day}^{-1/2}\cdot\text{share}^{-1}$, i.e., an annual volatility approximately equal to 21%,

- $V = 4{,}000{,}000 \ \text{shares}\cdot\text{day}^{-1}$,

- $L(\rho) = 0.1|\rho|^{0.75} + 0.004|\rho|$.

The call we consider is an at-the-money call ($K = 45$) with a maturity T of 63 days. Its nominal represents 5 times the daily market volume, i.e., $Q = 20{,}000{,}000$ shares. We also assume that the initial hedging portfolio is made of $q_0 = 0.5Q$ shares, that is, the initial Bachelier Δ.

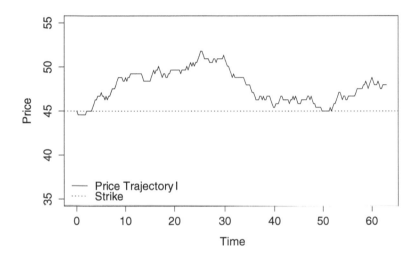

FIGURE 9.3: Stock price trajectory I.

Figure 9.3 represents a trajectory of the stock price, for which the option expires in the money ($S_T > K$). We see in Figure 9.4 the corresponding optimal hedging strategy, along with the classical[14] Bachelier Δ-hedging strategy.[15]

---

[14]The Δ in the Bachelier model is

$$\Delta = \Phi\left(\frac{S - K}{\sigma\sqrt{T - t}}\right),$$

where Φ is the cumulative distribution function of a Gaussian variable with mean 0 and variance 1.

[15]We have set $\rho_{\max}$ to a high value, so that the maximum participation constraint is never binding over the time interval $[0, T]$. We have also considered a risk-liquidity premium corresponding to a liquidation at a constant participation rate equal to 30%.

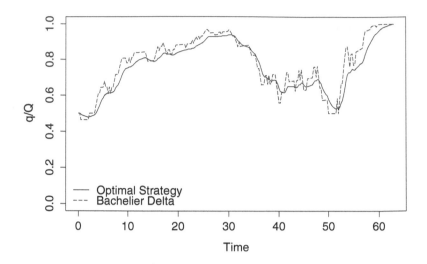

FIGURE 9.4: Optimal hedging trajectory and Bachelier Δ-hedging strategy I, for $\gamma = 10^{-7}$ €$^{-1}$.

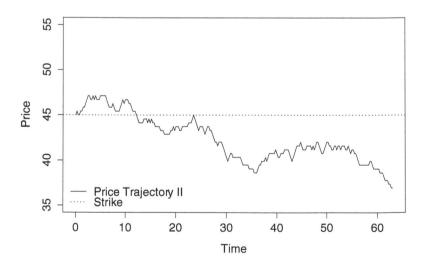

FIGURE 9.5: Stock price trajectory II.

Figure 9.5 represents a trajectory of the stock price for which the option expires worthless ($S_T < K$). We see in Figure 9.6 the corresponding optimal hedging strategy, along with the classical Bachelier Δ-hedging strategy.

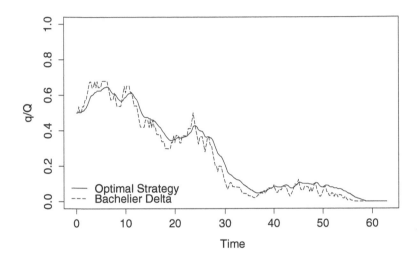

FIGURE 9.6: Optimal hedging trajectory and Bachelier Δ-hedging strategy II, for $\gamma = 10^{-7} \, \text{€}^{-1}$.

As expected, the hedging portfolio at time T contains Q shares in the case of trajectory I (the number of shares to deliver), whereas it contains none in the case of trajectory II. However, if the stock price fluctuates around the strike just before time T, then the portfolio will not contain either 0 or Q shares at time T, because of the limited available liquidity (hence the risk-liquidity premium at the end).

We also see in Figures 9.4 and 9.6 that the optimal hedging strategy in our model is not mean-reverting around the classical Bachelier Δ-hedging strategy, as in the model of Rogers and Singh [158] for instance. Nevertheless, the Bachelier Δ-hedging strategy is obtained as the limit of our model when the asset is very liquid (see Figure 9.7). Liquidity is important in our model both for the hedging strategy and the price of the option – see Table 9.2 for option prices (all prices are normalized by Q).

TABLE 9.2: Prices of the contract for different values of the liquidity parameter η ($\gamma = 10^{-7} \, \text{€}^{-1}$).

η	0.2	0.1	0.05	0.01	0 (Bachelier)
Price of the call	2.08	2.03	2.00	1.96	1.90

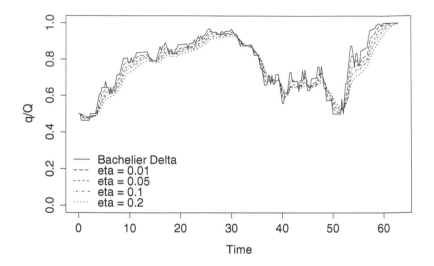

FIGURE 9.7: Optimal hedging trajectory for different values of the liquidity parameter η ($\gamma = 10^{-7}\ \text{€}^{-1}$).

Hedging strategies are smoother in our model than in the Bachelier model, and closer to the neutral position $0.5Q$. This is related to risk aversion. In fact, there are two risks of different natures:

- The first risk is related to the fluctuations in the price of the underlying. When we talk about hedging risk, we often think of this first risk.

- The second risk is related to the optional nature of the contract. The trader has to deliver either 0 or Q shares, and he will only know in which case he stands at time T. Because he is risk averse, the trader has an incentive to stay close to a neutral portfolio with $q = 0.5Q$ shares.

The above point regarding the shape of the hedging strategies is linked to the second risk. It is noteworthy that this second risk is never tackled in the literature.

It is also interesting to notice that the optimal hedging strategy depends a lot on the risk aversion parameter γ. In Figures 9.8 and 9.9, we see that, when γ is below our reference value of $10^{-7}\ \text{€}^{-1}$, the hedging strategy is very smooth, and the larger the γ, the more it fluctuates and the more it is close to the neutral portfolio $q = 0.5Q$.

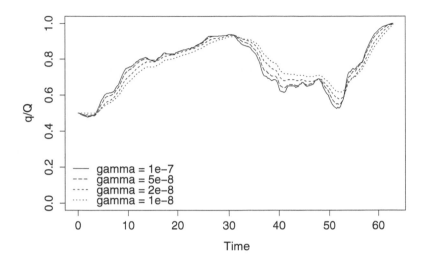

FIGURE 9.8: Optimal hedging trajectory for different values of the risk aversion parameter γ (1).

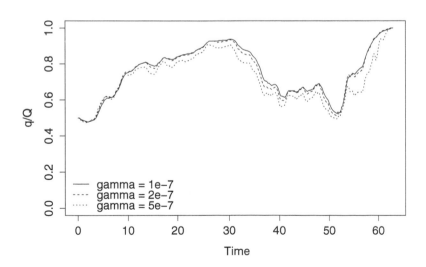

FIGURE 9.9: Optimal hedging trajectory for different values of the risk aversion parameter γ (2).

In terms of pricing, the level of risk aversion is also very important. Unsurprisingly, the price of the contract is an increasing function of γ (see Table 9.3).

TABLE 9.3: Price of the call option for different values of the risk aversion parameter γ.

γ (€^{-1})	10^{-8}	$2 \cdot 10^{-8}$	$5 \cdot 10^{-8}$	10^{-7}	$2 \cdot 10^{-7}$	$5 \cdot 10^{-7}$
Price of the call	1.95	1.96	1.99	2.03	2.09	2.29

To better understand what makes the model of this chapter different from the classical ones, it is interesting to discuss the influence of the initial portfolio. We see in Figure 9.10 that if we consider an empty hedging portfolio at the start, then it takes a few days before the hedging portfolio coincides with the portfolio of Figure 9.4. This effect is magnified when one sets an upper bound to the participation rate – see Figure 9.11.

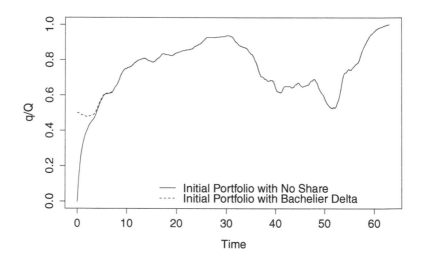

FIGURE 9.10: Optimal hedging trajectory for different values of q_0, for $\gamma = 10^{-7}\ \text{€}^{-1}$.

As far as prices are concerned, the (normalized) price of the option contract is 2.03 € when one starts with the Bachelier Δ-hedging portfolio, and 2.13 € when one starts with an empty portfolio. If one adds the constraint $\rho_{\max} = 50\%$, then the difference is even larger: 2.05 € when one starts with the Bachelier Δ-hedging portfolio, and 2.27 € when one starts with an empty portfolio.

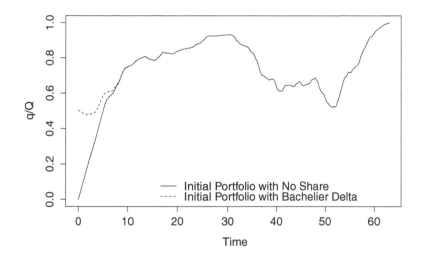

FIGURE 9.11: Optimal hedging trajectory for different values of q_0, when $\rho_{\max} = 50\%$ ($\gamma = 10^{-7}$ €$^{-1}$).

We have considered the case of a call option with physical delivery (and without permanent market impact). Through this example, we aimed at showing the outcome of an option pricing and hedging model built upon the Almgren-Chriss framework. For a detailed discussion on the difference between physical delivery and cash settlement, the influence of permanent market impact, and a precise comparison with classical Δ-hedging, we refer to the paper of Guéant and Pu [89].

9.5   Conclusion

Execution issues are omnipresent in finance. The hedging of equity derivatives is just a single example of the applications of execution models, outside of the cash equity world. In addition to its theoretical interest, we believe that the model presented in this chapter should be used in the case of options with huge nominal managed on a stand-alone basis, for instance in the framework of corporate deals.

Similar models can be used to tackle almost all problems in which liquidity matters. An interesting example is treated in the next chapter: that of share buy-back programs. Although many firms rely on open-market repurchase programs, many of them also use complex contracts to buy back shares: for instance, Accelerated Share Repurchase (ASR) contracts. We will see in Chapter 10 that ASRs raise both a classical execution problem and an option hedging problem (in fact the option at stake is a Bermudan option with an Asian-like payoff), and that these two problems should not be considered independently from each other. Therefore, the models presented in this book are perfectly suited for addressing the management of ASR contracts.

Chapter 10

Share buy-back

Nisi impunitatis cupido retinuisset, magnis semper conatibus adversa.

— Tacitus

In Chapter 9, we have built an option pricing and hedging model on top of an optimal execution model. This new model has made it possible to take account of the limited available liquidity and the market impact for building a better hedging strategy than the naïve Δ-hedging one. In this chapter, we go further: we build a model for a specific type of contracts that are at the same time execution contracts and option contracts. These contracts, called Accelerated Share Repurchase (ASR) contracts, are commonly used by listed firms to buy back their own shares. The management of these contracts should be done in practice with a model such as the one we present in this chapter, not with the classical risk-neutral pricing approach.

10.1 Introduction

10.1.1 Accelerated Share Repurchase contracts

Be it for taking advantage of stock undervaluation or in order to distribute part of their profits to shareholders, firms may buy back their own shares. The traditional way for a firm to repurchase its own shares is through open-market repurchase (OMR) programs.[1] However, as reported in [17], once they have announced their intention to buy back shares, many firms do not commit to their initial plan. Unexpected shocks on prices or on the liquidity of the stock may indeed provide incentives to slow down, postpone, or even cancel repurchase programs.

In order to make a credible commitment to buy back shares, more and more firms enter ASR contracts with investment banks. There are various

---

[1]Privately-negotiated repurchases and self-tenders represent a small percentage of the total amount of repurchases – see [49].

kinds of ASR contracts: we traditionally distinguish ASRs with fixed number of shares from fixed notional ASRs.

In the case of an ASR with fixed number of shares (Q), the contract is the following:[2]

1. At time $t = 0$, the bank borrows Q shares from shareholders (usually large institutional investors) and gives the shares to the firm in exchange of a fixed amount QS_0 where S_0 is the Mark-to-Market price of the stock at time $t = 0$. The bank then has to progressively buy back Q shares on the market to give back Q shares to the initial shareholders, and go from a short position to a flat position on the stock;

2. The bank is long an option with payoff $Q(A_\tau - S_0)$ (the firm being short of this option), where A_t is the average price between 0 and t (in practice the average of closing prices or the average of daily VWAPs), and where τ is chosen by the bank, among a set of specified dates.[3]

Therefore, the firm eventually pays (on average) a price A_τ for the Q shares that have been repurchased.

In the case of a fixed notional ASR with notional F, the contract is the following:[4]

1. At time $t = 0$, the bank borrows Q shares from shareholders – usually $Q = \alpha \frac{F}{S_0}$, where $\alpha \in [0, 1]$ is usually around 80% – and gives the shares to the firm in exchange of the fixed amount F;

2. An option is embedded in the contract; when the bank exercises this option at date τ (the date is chosen by the bank, among a set of specified dates), there is a transfer of $\frac{F}{A_\tau} - Q$ shares from the bank to the firm,[5] so that the actual number of shares obtained by the firm is $\frac{F}{A_\tau}$.

Therefore, the firm pays F and eventually gets $\frac{F}{A_\tau}$ shares.

As above, the bank will also progressively buy shares on the market to give back Q shares to the initial shareholders.

---

[2]We present here the case of a pre-paid ASR with fixed number of shares. We do not detail the numerous technical clauses that usually appear in ASR contracts.

[3]There is usually a latency period of a few weeks – after $t = 0$ – during which the option cannot be exercised.

[4]As above, we only discuss pre-paid contracts. Post-paid contracts also exist.

[5]α is chosen small enough so that the transfer is almost always from the bank to the firm, i.e., $\frac{F}{A_\tau} \geq Q$. In what follows, we implicitly assume that the settlement is from the bank to the firm.

10.1.2 Nature of the problem

Pricing these contracts and optimally managing them constitute mathematical problems that cannot be solved in a satisfactory way using the classical theory of derivatives pricing. At time $t = 0$, the bank has a short position on the stock and it will buy shares on the market over a period of time ranging from a few weeks to a few months. As a consequence, the problem is at the same time a problem of optimal execution and a problem of option pricing: a cash-settled exotic (Asian) option with Bermudan exercise dates in the case of an ASR with a fixed number of shares, and a physically settled exotic (Asian-like) option with Bermudan exercise dates in the case of a fixed notional ASR. Ignoring the interactions between the two problems would lead to a suboptimal strategy. Moreover, since the nominal/notional of these options is large, execution costs must be taken into account.

In what follows, we consider a discrete-time model – similar to the one developed in Chapter 9 for option pricing – in order to optimally manage ASR contracts. The model we present generalizes the one presented in Guéant, Pu, and Royer [90] for ASRs with fixed number of shares. Jaimungal, Kinzebulatov, and Rubisov [113] considered another model in continuous time, but only dealt with ASRs with fixed number of shares. We focus on the optimal management of both kinds of ASR contracts, that is, on the optimal execution strategy, and on the optimal exercise time. As far as pricing is concerned, we could discuss indifference prices for ASR contracts, as we did for options in Chapter 9. However, these contracts are often priced with a discount on the average price in the definition of the payoff of the contracts, and this makes the problem more complex (although the same methods can be used, at least numerically) – see [90].

10.2 The model

10.2.1 Setup of the model

The model we consider is a discrete-time model where each period of time (of length Δt) corresponds to one day. In other words, if the interval $[0, T]$ corresponds to N days, the time subdivision is $t_0 = 0 < \ldots < t_n = n\Delta t < \ldots < t_N = N\Delta t = T$.[6]

At the beginning of each time interval $[t_n, t_{n+1}]$, the bank chooses the number of shares it will buy (or sell) over this time interval. We denote by

---

[6]The model can be extended to model the overnight risk, but we do not consider this extension here.

$v_n \Delta t$ the number of shares bought by the bank between t_n and t_{n+1} (if this number is negative, it corresponds to selling shares instead of buying).

The dynamics for the number of shares bought by the bank is therefore

$$q_{n+1} = q_n + v_n \Delta t, \quad 0 \leq n < N.$$

Unlike the models we presented in the previous chapters, we assume here that q_n is not the position of the bank, but instead the number of shares bought by the bank. In particular, we assume $q_0 = 0$, whereas the initial position of the bank is a short one.

The price of the stock has the following dynamics:

$$S_{n+1} = S_n + \sigma \sqrt{\Delta t} \epsilon_{n+1}, \quad 0 \leq n < N,$$

where $\epsilon_1, \ldots, \epsilon_N$ are i.i.d. random variables with mean 0 and variance 1.

In particular, we assume that there is no permanent market impact. This may seem odd because the number of shares in buy-back contracts is usually very large – up to a few percents of the market capitalization. However, since the number of shares to buy back (or the dollar amount in the case of a fixed notional ASR) is usually public, we can consider that the market impact of the overall share buy-back contract precedes the actual transactions: it is already included in the market price of the stock, prior to the execution process. In other words, we assume that there is only a short-lasting market impact, i.e., execution costs.

The average price process in the definition of the option payoff in ASR contracts is denoted by $(A_n)_{n \geq 1}$, where[7]

$$A_n = \frac{1}{n} \sum_{k=0}^{n-1} S_{k+1}.$$

The dynamics of the cash account is

$$X_{n+1} = X_n - v_n S_{n+1} \Delta t - L\left(\frac{v_n}{V_{n+1}}\right) V_{n+1} \Delta t, \quad 0 \leq n < N,$$

where $V_{n+1} \Delta t$ is the market volume over $[t_n, t_{n+1}]$,[8] and where L is an execution cost function, as in Chapter 3.

---

[7]Depending on the exact definition of the payoff of the option embedded in the ASR contract, we assume therefore that S_{n+1} is either the closing price at the end of the time interval $[t_n, t_{n+1}]$, or the VWAP over this time interval.

[8]As above, the market volume process is assumed to be deterministic.

As opposed to what we assumed in Chapter 9, the execution costs paid to buy or sell shares are here relative to S_{n+1}. This is because banks executing ASR contracts tend to use VWAP orders or Target Close orders, in line with the definition of $(A_n)_n$.[9]

We then introduce a nonempty set $\mathcal{N} \subset \{1, \ldots, N-1\}$ corresponding to time indices at which the bank can choose to exercise the option (notice that $N \notin \mathcal{N}$, because the bank has no choice at time $t_N = T$: the settlement occurs at time $t_N = T$ if it has not occurred beforehand). In practice, this set is usually of the form $\{n_0, \ldots, N-1\}$, where $n_0 > 1$.

When the bank decides to exercise the option, or at maturity if the option has not been exercised yet, the settlement of the contract occurs. We denote by $n^\star \in \mathcal{N} \cup \{N\}$ the time index corresponding to the settlement.

In the case of an ASR with a fixed number of shares Q, the bank receives at time $t_{n^\star}$ the amount $Q(A_{n^\star} - S_0)$, in addition to the upfront payment QS_0. Moreover, the bank has bought $q_{n^\star}$ shares and $Q - q_{n^\star}$ shares still need to be bought after time $t_{n^\star}$ to give them back to the initial shareholders. Because we consider a CARA utility function, and because the latter execution problem, after time $t_{n^\star}$, only depends on $\mathcal{F}_{n^\star}$ through $q_{n^\star}$, we can consider a ($\mathcal{F}_{n^\star}$-measurable) certainty equivalent for the latter execution problem. In other words, we assume that the $Q - q_{n^\star}$ shares are bought for a cash amount equal to

$$S_{n^\star}(Q - q_{n^\star}) + \ell(Q - q_{n^\star}),$$

where $\ell(\cdot)$ is a risk-liquidity premium function (see Chapter 8).[10]

The problem we consider is therefore to maximize the expression

$$
\begin{aligned}
&\mathbb{E}\left[-\exp\left(-\gamma\left(X_{n^\star} + QA_{n^\star} - S_{n^\star}(Q - q_{n^\star}) - \ell(Q - q_{n^\star})\right)\right)\right] \\
=\ &\mathbb{E}\left[-\exp\left(-\gamma\left(X_{n^\star} + q_{n^\star}S_{n^\star} + Q(A_{n^\star} - S_{n^\star}) - \ell(Q - q_{n^\star})\right)\right)\right]
\end{aligned}
$$

over the set of admissible controls[11]

$$
\begin{aligned}
\mathcal{A} = \Big\{ (v, n^\star) \Big| \quad &v = (v_n)_{0 \le n \le n^\star - 1} \text{ is } (\mathcal{F})\text{-adapted}, |v_n| \le \rho_{\max} V_{n+1}, \\
&n^\star \text{ is a } (\mathcal{F})\text{-stopping time taking values in } \mathcal{N} \cup \{N\} \Big\}.
\end{aligned}
$$

In the case of an ASR with fixed notional F, the bank has received F in cash at time $t = 0$. At time $t_{n^\star}$, the bank has bought $q_{n^\star}$ shares on the market and delivered $\frac{F}{A_{n^\star}} - Q$ shares to the firm. Furthermore, it needs to continue

---

[9]The price risk over the day can be modeled by a premium applied to the execution function L.

[10]To be absolutely rigorous, here it should be the risk-liquidity premium function obtained in the discrete-time model.

[11]We consider here a maximum constraint on the participation rate.

buying back shares in order to give Q shares back to the initial shareholders. Summing up, $\frac{F}{A_{n^*}} - q_{n^*}$ shares still need to be bought after time t_{n^*}.

As above, we can consider a ($\mathcal{F}_{n^*}$-measurable) certainty equivalent for the latter execution problem, and assume that the $\frac{F}{A_{n^*}} - q_{n^*}$ shares are bought for a cash amount equal to

$$
S_{n^*}\left(\frac{F}{A_{n^*}} - q_{n^*}\right) + \ell\left(\frac{F}{A_{n^*}} - q_{n^*}\right).
$$

The problem then boils down to maximizing

$$
\mathbb{E}\left[-\exp\left(-\gamma\left(X_{n^*} + F - S_{n^*}\left(\frac{F}{A_{n^*}} - q_{n^*}\right) - \ell\left(\frac{F}{A_{n^*}} - q_{n^*}\right)\right)\right)\right]
$$
$$
= \mathbb{E}\left[-\exp\left(-\gamma\left(X_{n^*} + q_{n^*}S_{n^*} + \frac{F}{A_{n^*}}(A_{n^*} - S_{n^*}) - \ell\left(\frac{F}{A_{n^*}} - q_{n^*}\right)\right)\right)\right],
$$

over the same set $\mathcal{A}$ of admissible controls as above.

ASRs with fixed number of shares and fixed notional ASRs can therefore be considered in the same modeling framework if we consider the maximization over $\mathcal{A}$ of

$$
\mathbb{E}\left[-\exp\left(-\gamma\left(X_{n^*} + q_{n^*}S_{n^*} + \Pi(q_{n^*}, S_{n^*}, A_{n^*})\right)\right)\right],
$$

where

$$
\Pi(q, S, A) = \begin{cases} Q(A - S) - \ell(Q - q) & \text{in the former case,} \\ \frac{F}{A}(A - S) - \ell\left(\frac{F}{A} - q\right) & \text{in the latter case.} \end{cases} \tag{10.1}
$$

10.2.2   Towards a recursive characterization of the optimal strategy

Solving the above problem requires to determine both the optimal execution strategy of the bank and the optimal exercise time. For that purpose, we use dynamic programming. We introduce the value functions $(u_n)_{0 \le n \le N}$ defined by[12]

$$
u^n(x, q, S, A) =
$$
$$
\sup_{(v,n^*)\in\mathcal{A}_n}\mathbb{E}\left[-\exp\left(-\gamma\left(X_{n^*}^{n,x,v} + q_{n^*}^{n,q,v}S_{n^*}^{n,S} + \Pi(q_{n^*}^{n,q,v}, S_{n^*}^{n,S}, A_{n^*}^{n,A,S})\right)\right)\right],
$$

where $\mathcal{A}_n$ is the set of admissible strategies at time t_n defined by

$$
\mathcal{A}_n = \left\{(v, n^*) \;\middle|\; \begin{array}{l} v = (v_k)_{n \le k \le n^*-1} \text{ is } (\mathcal{F})\text{-adapted}, |v_k| \le \rho_{\max}V_{k+1}, \\ n^* \text{ is a } (\mathcal{F})\text{-stopping time taking} \\ \text{values in } (\mathcal{N} \cup \{N\}) \cap \{n, \ldots, N\} \end{array}\right\},
$$

---

[12]In fact, when $n = 0$, the function u_n is independent of A.

and where the state variables are defined for $0 \leq n \leq k \leq N$ by

$$X_k^{n,x,v} = x - \sum_{j=n}^{k-1} \left(v_j S_{j+1}^{n,S} \Delta t + L\left(\frac{v_j}{V_{j+1}}\right) V_{j+1} \Delta t \right), \qquad (10.2)$$

$$q_k^{n,q,v} = q + \sum_{j=n}^{k-1} v_j \Delta t, \qquad (10.3)$$

$$S_k^{n,S} = S + \sigma\sqrt{\Delta t} \sum_{j=n}^{k-1} \epsilon_{j+1}, \qquad (10.4)$$

$$A_k^{n,A,S} = \frac{n}{k} A + \frac{1}{k} \sum_{j=n}^{k-1} S_{j+1}^{n,S}, \quad (k > 0). \qquad (10.5)$$

The family of functions $(u^n)_{0 \leq n \leq N}$ defined above is the solution of the Bellman equation:

$$u^n(x, q, S, A) = \begin{cases} -\exp\left(-\gamma\left(x + qS + \Pi(q, S, A)\right)\right) & \text{if } n = N, \\ \max\left\{ \tilde{u}_{n,n+1}\left(x, q, S, A\right), \right. & \\ \qquad \left. -\exp\left(-\gamma\left(x + qS + \Pi(q, S, A)\right)\right) \right\} & \text{if } n \in \mathcal{N}, \\ \tilde{u}_{n,n+1}\left(x, q, S, A\right) & \text{otherwise}, \end{cases}$$

where

$$\tilde{u}_{n,n+1}(x, q, S, A) = \sup_{|v| \leq \rho_{\max} V_{n+1}} \mathbb{E}\left[u^{n+1}\left(X_{n+1}^{n,x,v}, q_{n+1}^{n,q,v}, S_{n+1}^{n,S}, A_{n+1}^{n,A,S} \right) \right].$$

Using the same reasoning as in Chapter 9, we can factor out the term $x + qS$. In other words, we write the value functions as

$$u^n(x, q, S, A) = -\exp\left(-\gamma\left(x + qS - \theta^n\left(q, S, A\right)\right)\right). \qquad (10.6)$$

It is straightforward to verify that the functions $(\theta^n)_n$ are characterized by the following recursive equations:

$$\theta^n(q, S, A) = \begin{cases} -\Pi(q, S, A) & \text{if } n = N, \\ \min\left\{ \tilde{\theta}_{n,n+1}\left(q, S, A\right), -\Pi(q, S, A) \right\} & \text{if } n \in \mathcal{N}, \\ \tilde{\theta}_{n,n+1}\left(q, S, A\right) & \text{otherwise}, \end{cases}$$

$$(10.7)$$

where

$$\tilde{\theta}_{n,n+1}(q, S, A) =$$

$$\frac{1}{\gamma} \inf_{|v| \leq \rho_{\max} V_{n+1}} \log \mathbb{E}\left[\exp\left(\gamma\left(L\left(\frac{v}{V_{n+1}}\right) V_{n+1}\Delta t - \sigma q\sqrt{\Delta t}\epsilon_{n+1} \right. \right.\right.$$

$$\left.\left.\left. + \theta^{n+1}\left(q + v\Delta t, S + \sigma\sqrt{\Delta t}\epsilon_{n+1}, \frac{n}{n+1}A + \frac{1}{n+1}(S + \sigma\sqrt{\Delta t}\epsilon_{n+1}) \right) \right) \right) \right].$$

$$(10.8)$$

These recursive equations deserve several remarks.

A first important remark is that Eq. (10.8) is very similar to Eq. (9.12). We have indeed in both cases a log-exp transform of a random variable that takes into account the present risk (this is the term $-\sigma q\sqrt{\Delta t}\epsilon_{n+1}$), the execution costs of the next trade (this is the term $L\left(\frac{v}{V_{n+1}}\right) V_{n+1}\Delta t$), and the future state of the problem, once the next trade is done. The differences with the model of Chapter 9 are due to: (i) the fact that execution costs over the time interval $[t_n, t_{n+1}]$ are measured with respect to the price S_{n+1} and not with respect to the price S_n, and (ii) the role played by the average price A.[13]

Another remark is that because u_0 is independent of A, θ_0 is also independent of A. Therefore, $\tilde{\theta}_{0,1}$ should also be independent of A. Although this is not obvious at first sight, there is no dependence on A in Eq. (10.8) when $n = 0$ because $\frac{n}{n+1}A$ is equal to 0.

The change of variables given by Eq. (10.6) reduces the dimension of the problem from 5 to 4. The variable x does not appear anymore indeed in the functions $(\theta_n)_n$. In Jaimungal et al. [113], and similarly in Guéant et al. [90], the dimension of the problem goes from 5 to 3. In the case of ASR contracts with fixed number of shares, it can indeed be shown that both the optimal trading decisions and the optimal exercise time depend on the couple (S, A) in a particular manner. In the case of [113], the relevant variable is indeed $\frac{A}{S}$. In the case of [90], the relevant variable is instead the spread $A - S$. This reduction of dimension from 5 to 3 is however no longer possible in the case of fixed notional ASRs. The additional convexity in the payoff (the term $\frac{F}{A}$) leads to a problem of dimension 4, not 3.

---

[13]Of course, another difference is that Eq. (10.8) is involved in Eq. (10.7), which determines when it is optimal to exercise the option.

10.3 Optimal management of an ASR contract

10.3.1 Characterization of the optimal trading strategy and the optimal exercise time

An important point to notice is that Eqs. (10.7) and (10.8) make it possible to find the optimal trading strategy and to make the optimal decision with respect to the exercise time:

- The optimal number of shares $v_n^*(q, S, A)$ to buy over $[t_n, t_{n+1}]$ is given by

$$\operatorname{argmin}_{|v| \leq \rho_{\max} V_{n+1}} \log \mathbb{E} \left[\exp \left(\gamma \left(L \left(\frac{v}{V_{n+1}} \right) V_{n+1} \Delta t - \sigma q \sqrt{\Delta t} \epsilon_{n+1} \right. \right. \right.$$

$$\left. \left. \left. + \theta^{n+1} \left(q + v \Delta t, S + \sigma \sqrt{\Delta t} \epsilon_{n+1}, \frac{n}{n+1} A + \frac{1}{n+1} (S + \sigma \sqrt{\Delta t} \epsilon_{n+1}) \right) \right) \right) \right].$$

- The decision to exercise the option (for $n \in \mathcal{N}$) depends on the comparison between $\tilde{\theta}_{n,n+1}(q, S, A)$ and $-\Pi(q, S, A)$. If the latter is smaller than the former, then it is optimal to exercise the option.[14]

10.3.2 Analysis of the optimal behavior

To better understand the different drivers of the optimal strategy, it is interesting to write down the definition of the function θ_n. By using Eq. (10.6), we have indeed that

$$\theta^n(q, S, A) = x + qS$$

$$+ \inf_{(v,n^*) \in \mathcal{A}_n} \frac{1}{\gamma} \log \mathbb{E} \left[\exp \left(-\gamma \left(X_{n^*}^{n,x,v} + q_{n^*}^{n,q,v} S_{n^*}^{n,S} + \Pi(q_{n^*}^{n,q,v}, S_{n^*}^{n,S}, A_{n^*}^{n,A,S}) \right) \right) \right].$$

By using Eqs. (10.2), (10.3), and (10.4), we know that

$$X_{n^*}^{n,x,v} + q_{n^*}^{n,q,v} S_{n^*}^{n,S} = x + qS - \sum_{j=n}^{n^*-1} L \left(\frac{v_j}{V_{j+1}} \right) V_{j+1} \Delta t + \sigma \sqrt{\Delta t} \sum_{j=n}^{n^*-1} q_j \epsilon_{j+1}.$$

---

[14]The model does not take into account the fact that trading decisions are not taken at the same time as the decision to exercise or not, but this is a second-order problem.

Therefore, we have

$$\theta^n(q, S, A) =$$

$$\inf_{(v,n^\star)\in\mathcal{A}_n} \frac{1}{\gamma} \log \mathbb{E}\left[\exp\left(-\gamma\left(-\sum_{j=n}^{n^\star-1} L\left(\frac{v_j}{V_{j+1}}\right)V_{j+1}\Delta t + \sigma\sqrt{\Delta t}\sum_{j=n}^{n^\star-1} q_j\epsilon_{j+1}\right.\right.\right.$$

$$\left.\left.\left.+\Pi(q_{n^\star}^{n,q,v}, S_{n^\star}^{n,S}, A_{n^\star}^{n,A,S})\right)\right)\right].$$

Now, we need to expand $\Pi(q_{n^\star}^{n,q,v}, S_{n^\star}^{n,S}, A_{n^\star}^{n,A,S})$. We see from Eq. (10.1) that for both kinds of ASRs, we need to compute $A_{n^\star}^{n,A,S} - S_{n^\star}^{n,S}$.

From Eqs. (10.4) and (10.5), we have

$$
\begin{aligned}
A_{n^\star}^{n,A,S} - S_{n^\star}^{n,S} &= \frac{n}{n^\star}A + \frac{1}{n^\star}\left(\sum_{m=n}^{n^\star-1} S_{m+1}^{n,S}\right) - S_{n^\star}^{n,S} \\
&= \frac{n}{n^\star}A + \frac{n^\star - n}{n^\star}S + \frac{1}{n^\star}\sigma\sqrt{\Delta t}\left(\sum_{m=n}^{n^\star-1}\left(\sum_{j=n}^{m}\epsilon_{j+1}\right)\right) - S_{n^\star}^{n,S} \\
&= \frac{n}{n^\star}(A - S) + S + \frac{1}{n^\star}\sigma\sqrt{\Delta t}\left(\sum_{j=n}^{n^\star-1}(n^\star - j)\epsilon_{j+1}\right) - S_{n^\star}^{n,S} \\
&= \frac{n}{n^\star}(A - S) - \frac{1}{n^\star}\sigma\sqrt{\Delta t}\sum_{j=n}^{n^\star-1} j\epsilon_{j+1}.
\end{aligned}
$$

In the case of an ASR with a fixed number of shares Q, we have therefore[15]

$$\theta^n(q, S, A) =$$

$$\inf_{(v,n^\star)\in\mathcal{A}_n} \frac{1}{\gamma} \log \mathbb{E}\left[\exp\left(-\gamma\left(-\sum_{j=n}^{n^\star-1} L\left(\frac{v_j}{V_{j+1}}\right)V_{j+1}\Delta t\right.\right.\right.$$

$$\left.\left.\left.+\sigma\sqrt{\Delta t}\sum_{j=n}^{n^\star-1}\left(q_j - \frac{j}{n^\star}Q\right)\epsilon_{j+1} + \frac{n}{n^\star}Q(A - S) - \ell(Q - q_{n^\star}^{n,q,v})\right)\right)\right]. \quad (10.9)$$

---

[15]We see in particular that $\theta^n(q, S, A)$ is in fact a function of q and $A - S$, as in [90].

In the case of an ASR with fixed notional F, we have

$$\theta^n(q, S, A) = \tag{10.10}$$

$$\inf_{(v,n^*)\in\mathcal{A}_n} \frac{1}{\gamma} \log \mathbb{E}\left[\exp\left(-\gamma\left(-\sum_{j=n}^{n^*-1} L\left(\frac{v_j}{V_{j+1}}\right) V_{j+1}\Delta t \right.\right.\right.$$

$$\left.\left.\left. +\sigma\sqrt{\Delta t} \sum_{j=n}^{n^*-1} \left(q_j - \frac{j}{n^*}\frac{F}{A_{n^*}}\right)\epsilon_{j+1} + \frac{n}{n^*}\frac{F}{A_{n^*}}(A - S) - \ell\left(\frac{F}{A_{n^*}} - q_{n^*}^{n,q,v}\right) \right) \right) \right].$$

To understand the different effects, one has to analyze what happens before and after the exercise date of the option.

Before time t_{n^*}, the bank buys back shares on the market. As for all execution problems, there is an execution cost component and a risk component:

- The bank pays execution costs as a function of its participation rate. Hence the term

$$\sum_{j=n}^{n^*-1} L\left(\frac{v_j}{V_{j+1}}\right) V_{j+1}\Delta t$$

in both Eq. (10.9) and Eq. (10.10).

- The bank is exposed to price moves. However, due to the form of the payoff, the risk is partially hedged. When the price goes up, the bank pays more to buy back shares, but the average price entering the definition of the option payoff also increases. The risk is measured by the term

$$\sigma\sqrt{\Delta t} \sum_{j=n}^{n^*-1} \left(q_j - \frac{j}{n^*}Q\right)\epsilon_{j+1},$$

in the case of an ASR with fixed number of shares, and by the term

$$\sigma\sqrt{\Delta t} \sum_{j=n}^{n^*-1} \left(q_j - \frac{j}{n^*}\frac{F}{A_{n^*}}\right)\epsilon_{j+1},$$

in the case of a fixed notional ASR. In particular, the risk can be completely hedged away in the former case if one decides to set $n^* = N$ and to trade at a constant pace.[16]

After time t_{n^*}, the execution costs are similar but the risk is no longer partially hedged. Once the option has been exercised, the problem becomes indeed a classical execution problem, and this is the reason why execution

---

[16]This is not the optimal strategy, because it reduces the value of the option to 0.

costs and price risk are together modeled by a risk-liquidity premium; hence the term

$$\ell(Q - q_{n*}^{n,q,v})$$

in Eq. (10.9), and the term

$$\ell\left(\frac{F}{A_{n*}} - q_{n*}^{n,q,v}\right)$$

in Eq. (10.10).

The above analysis is linked to the execution process in order to buy back shares. As far as the optimal stopping problem is concerned, the lower the price S compared to A, the more the bank has an incentive to exercise the option. The cost of not exercising now is somehow given by the spread terms $\frac{n}{n*}Q(A - S)$ and $\frac{n}{n*}\frac{F}{A_{n*}}(A - S)$ of Eqs. (10.9) and (10.10).

We see therefore that there are two effects at play. First, there is an incentive to exercise the option when S goes below A; in such a situation, one buys shares at a low price and gets paid A in the case of an ASR with fixed number of shares (or delivers less shares in the other case). Second, there is an incentive not to exercise too early in order to (partially) hedge the risk associated with the execution process. Overall, the optimal strategy consists therefore in accelerating the buying process when S decreases, in order to have less to buy without the natural hedge of the option if it turns out to be optimal to exercise the option, and decelerating the buying process (or even selling shares as we will see below) when S increases, to go to an execution trajectory that highly benefits from the hedging power of the option component. The option is then exercised when the difference between A and S brings a profit that compensates the additional risk of the execution process after the exercise date of the option.

10.4 Numerical methods and examples

10.4.1 A pentanomial-tree approach

To solve the problem numerically, we approximate the functions $(\theta^n)_n$ by backward induction using Eqs. (10.7) and (10.8). The method we propose is a tree-based approach for the diffusion of prices.[17]

For each node in the tree, indexed by a time index n and a price S, we approximate the values of the function $\theta_n(\cdot, S, \cdot)$ on a (q, A)-grid.

---

[17]In the case of ASRs with fixed number of shares, one can directly diffuse $A - S$, see [90].

To build the tree, we consider that the $(\epsilon_n)_n$ are i.i.d. with the following distribution:[18]

$$\epsilon_n = \begin{cases} +2 & \text{with probability } \frac{1}{12}, \\ +1 & \text{with probability } \frac{1}{6}, \\ 0 & \text{with probability } \frac{1}{2}, \\ -1 & \text{with probability } \frac{1}{6}, \\ -2 & \text{with probability } \frac{1}{12}. \end{cases}$$

In other words, we build a recombinant pentanomial tree, with, at the time step n, $4n + 1$ nodes corresponding to the following set of prices:

$$\left\{ S_0 + k\sigma\sqrt{\Delta t}, -2n \le k \le 2n \right\}.$$

At each node (n, S) of the tree, we use Eqs. (10.7) and (10.8) to approximate the values of $\theta^n(q, S, A)$ for $(q, A) \in \mathcal{G}_q \times \mathcal{G}_A$, where $\mathcal{G}_q$ is a grid of the form

$$\left\{ \frac{k}{n_q - 1} q_{\max}, k \in \{0, \dots, n_q - 1\} \right\}$$

and where $\mathcal{G}_A$ is a grid of the form

$$\left\{ S_0 + \xi \left(\frac{k}{n_A - 1} - \frac{1}{2} \right) \sigma\sqrt{T}, k \in \{0, \dots, n_A - 1\} \right\}.$$

For each grid, we need to choose the number of points in the grid (n_q and n_A) and the width of the grid (linked to $q_{\max}$ and ξ).

Because $\mathbb{V}[A_N] \simeq \frac{1}{3}\sigma^2 T$, it is natural to consider values of A in a range of the form $\left[S_0 - \frac{\omega}{\sqrt{3}}\sigma\sqrt{T}, S_0 + \frac{\omega}{\sqrt{3}}\sigma\sqrt{T} \right]$, where ω is the number of standard deviations we want to consider. In the numerical examples below, we consider $\xi = 3$, that is, $\omega \simeq 2.6$ standard deviations for A_N.

The number of shares to deliver is either a constant Q, in the case of an ASR with a fixed number of shares, or a variable $\frac{F}{A}$, in the case of a fixed notional ASR. In the former case, it is natural to consider $q_{\max} = Q$. In the latter case, in coherence with the choice for A, we consider in the examples presented below a value $q_{\max} \simeq \frac{F}{S_0 - \frac{\xi}{2}\sigma\sqrt{T}}$.

Given the tree and the (q, A)-grid, we approximate $(\theta^n)_n$ by backward induction,[19] by using Eq. (10.7).

---

[18] Probabilities are chosen, so that the mean is 0, the variance 1, and the kurtosis 3.

[19] There is the same abuse of notation as in Chapter 9, because we use the same notation for both the function and its approximation.

We obtain

$$\tilde{\theta}_{n,n+1}(q,S,A) = \frac{1}{\gamma} \min_{\mathfrak{q}\in\mathcal{G}_q, |q-\mathfrak{q}|\le\rho_{\max}V_{n+1}\Delta t}$$

$$\log\mathbb{E}\left[\exp\left(\gamma\left(L\left(\frac{\mathfrak{q}-q}{V_{n+1}\Delta t}\right)V_{n+1}\Delta t - \sigma q\sqrt{\Delta t}\epsilon_{n+1}\right.\right.\right.$$

$$\left.\left.\left.+\theta^{n+1}\left(\mathfrak{q}, S+\sigma\sqrt{\Delta t}\epsilon_{n+1}, \frac{n}{n+1}A + \frac{1}{n+1}(S+\sigma\sqrt{\Delta t}\epsilon_{n+1})\right)\right)\right)\right].$$

To find the minimizer(s) in the above expression, and then compute $\tilde{\theta}_{n,n+1}(q,S,A)$, we need the values of

$$\theta^{n+1}\left(\mathfrak{q}, S+\sigma\sqrt{\Delta t}\epsilon_{n+1}, A'\right)$$

for $A' = \frac{n}{n+1}A + \frac{1}{n+1}(S+\sigma\sqrt{\Delta t}\epsilon_{n+1})$. As A' does not necessarily lie on the grid $\mathcal{G}_A$, we use interpolation with natural cubic splines and, if necessary, we extrapolate linearly the functions outside of the domain (for A).

10.4.2 Numerical examples

We now consider examples of ASR contracts. For that purpose, we consider a stock with the following characteristics (inspired from TOTAL SA):

- $S_0 = 45$ €.

- $\sigma = 0.6$ €·day$^{-1/2}$·share^{-1}, which corresponds to an annual volatility approximately equal to 21%.

- $V = 4{,}000{,}000$ stocks·day^{-1}.

- $L(\rho) = \eta|\rho|^{1+\phi} + \psi|\rho|$ with $\eta = 0.1$ €·stock^{-1}, $\psi = 0.004$ €·stock^{-1}, and $\phi = 0.75$.

For both kinds of ASRs, we consider that the terminal date corresponds to 3 months ($T = 63$ days). We also assume that the option cannot be exercised over the first month. In other words, the set of possible dates for an early exercise of the option is $\mathcal{N} = \{22,\ldots,62\}$.

As far as execution is concerned, we assume the following:

- The participation rate is bounded by $\rho_{\max} = 25\%$.

- ℓ corresponds to a liquidation at a constant participation rate of 25%. In other words, after the exercise date of the option, we assume that execution is done with a constant participation rate of 25%. To stick to the model, we should have assumed instead an IS strategy (with a maximal constraint $\rho_{\max} = 25\%$ on the participation rate), but there is little difference in terms of risk-liquidity premium.

For the examples we propose, we have chosen[20] $\gamma = 10^{-7}$ €$^{-1}$.

We consider an ASR contract with fixed number of shares $Q = 20{,}000{,}000$, and a fixed notional ASR with $F = 900{,}000{,}000$ €.[21]

The first example we consider (Figure 10.1) corresponds to a stock price mainly lying above the average price A. As explained in the previous section, there is no reason in that case to exercise the option early, because A increases (the dot corresponds to the date at which the option is exercised). We also see that the buying process accelerates when the stock price decreases, and conversely that it decelerates when the stock price increases. In particular, the bank can even sell shares. The rationale for the bank selling back shares is that it anticipates it will exercise late, therefore it is optimal to go back to a risk-free[22] straight-line trajectory corresponding to a late exercise date.

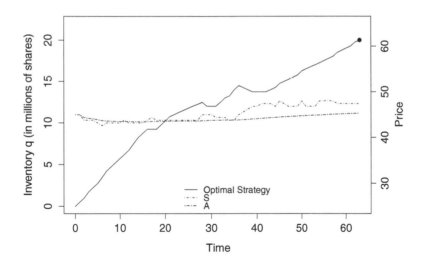

FIGURE 10.1: Optimal strategy for an ASR with fixed number of shares ($Q = 20{,}000{,}000$ shares) – Price Trajectory I.

---

[20] The influence of γ is discussed at the end of the chapter.
[21] We have chosen $F = QS_0$ to be able to compare the two kinds of ASRs.
[22] In the case of an ASR with fixed number of shares.

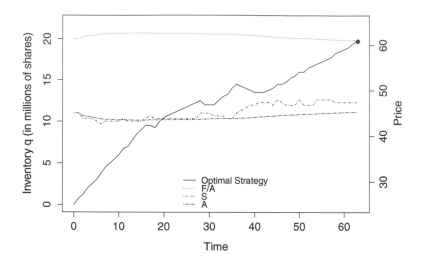

FIGURE 10.2: Optimal strategy for an ASR with fixed notional ($F =$ 900,000,000 €) – Price Trajectory I.

The same effects are seen in the case of the fixed notional ASR (see Figure 10.2). The only difference is that straight-line trajectories are not risk-free, but the above argument still works: the bank sells to go back to an execution trajectory where the risk of the execution process is partially hedged by the payoff of the option.

The second trajectory we consider corresponds to decreasing prices – at least over the first part of the period (see Figures 10.3 and 10.4). In that case, the bank buys shares quite rapidly and exercises the option as soon as it can, even though it has not yet bought the required number of shares. By exercising the option, the bank wants in fact to avoid the natural decrease in A that would lead to a decrease in its payoff; it is ready to pay fast-execution costs and to lose the natural hedge of the option payoff, in order to benefit from the difference between A and S.

The third trajectory (Figures 10.5 and 10.6) corresponds to a stock price oscillating around its average value. We see, as above, that the buying process accelerates when the stock price decreases. After 20 days, the bank would have exercised the option had it been allowed to do it. However, the price suddenly increased, and at the end of day 22, exercising was not optimal anymore. The buying process turned into a selling process to go back to an execution trajectory where the risk of the execution process is (partially) hedged by the option component. Then, when the price started to decrease again, the bank

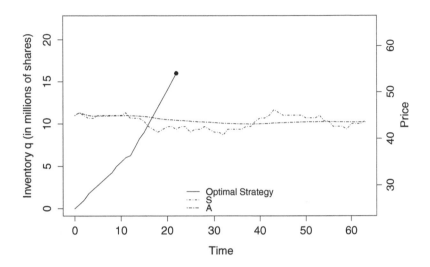

FIGURE 10.3: Optimal strategy for an ASR with fixed number of shares ($Q = 20,000,000$ shares) – Price Trajectory II.

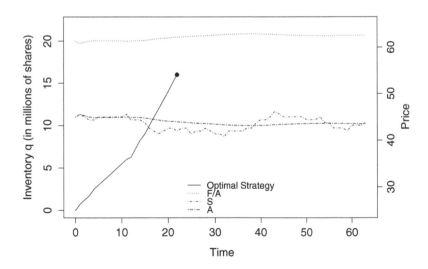

FIGURE 10.4: Optimal strategy for an ASR with fixed notional ($F = 900,000,000$ €) – Price Trajectory II.

accelerated its buying process, and eventually exercised the option soon after day 50. At that date, the difference between A and S was not that large, but it was worth exercising because there were only a few shares to buy back.

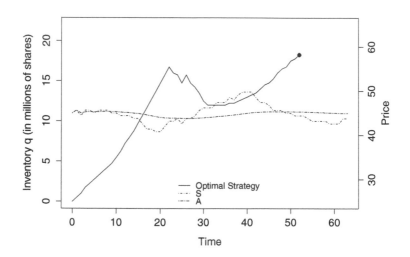

FIGURE 10.5: Optimal strategy for an ASR with fixed number of shares ($Q = 20,000,000$ shares) – Price Trajectory III.

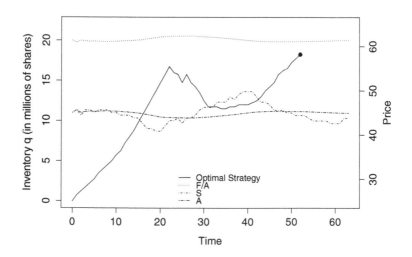

FIGURE 10.6: Optimal strategy for an ASR with fixed notional ($F = 900,000,000 €$) – Price Trajectory III.

The above examples deserve several remarks:

- We see that there is little difference between an ASR contract with fixed number of shares and a fixed notional ASR. If one computes indifference prices for these contracts, one notices that fixed notional ASR contracts are more profitable for the bank, but the difference is usually very small.

- We also see that the option component in the ASR contract constitutes a hedge for the execution process. Therefore, considering the execution problem and the option hedging problem separately would be a mistake.

- The optimal trajectory is highly influenced by the willingness to reduce risk. Therefore the risk aversion parameter plays an important role. This is instanced in Figures 10.7 and 10.8. We see that the choice of γ is crucial. We also see that, when $\gamma = 0$, the strategy is a buy-only strategy.

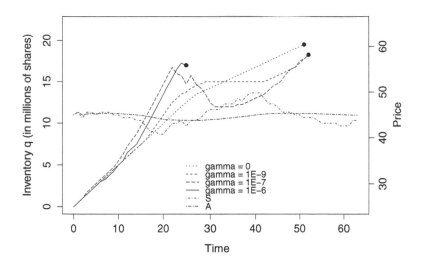

FIGURE 10.7: Optimal strategy for an ASR with fixed number of shares ($Q = 20,000,000$ shares) – Price Trajectory III – for different values of the risk aversion parameter γ.

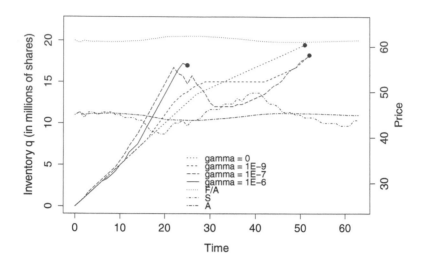

FIGURE 10.8: Optimal strategy for an ASR with fixed notional ($F = 900,000,000 \, \text{€}$) – Price Trajectory III – for different values of the risk aversion parameter γ.

10.5 Conclusion

For contracts that are at the same time execution contracts and option contracts, building a model on top of the Almgren-Chriss framework is a better choice than (blindly) using the risk-neutral pricing approach – for a comparison, see for instance [113]. We hope that this chapter on Accelerated Share Repurchase contracts has persuaded the reader of the interest of the Almgren-Chriss framework outside of pure execution problems. Other applications are promising, for instance to price collateralized loans, or in asset management. However, all the examples of Part III of this book have shown that the choice of a level of risk aversion was crucial. The latter point can be seen as a caveat for our approach, but we think that it is always better to have a degree of freedom.

Part IV

Market Making

Chapter 11

Market making models: from Avellaneda-Stoikov to Guéant-Lehalle, and beyond

Blofeld: "James Bond, allow me to introduce myself. I am Ernst Stavro
Blofeld. They told me you were assassinated in Hong Kong."
Bond: "Yes, this is my second life."

— *You Only Live Twice*

In the previous chapters, we mainly dealt with optimal execution: the optimal execution of a large order to liquidate a portfolio, the optimal execution of a hedging strategy, and the optimal execution strategy in the case of a buy-back contract. Therefore, our focus was mainly on liquidity taking issues – even though we also considered the use of liquidity-providing (limit) orders to buy or sell shares, for instance when we dealt with tactics in Chapter 7. In this chapter on market making, we focus instead on liquidity provision.

We first present the problem faced by market makers on order-driven and quote-driven markets. We then present the most famous model for market making, i.e., the Avellaneda-Stoikov model [13], together with the closed-form approximations of the optimal bid and ask quotes derived by Guéant, Lehalle, and Fernandez-Tapia [88]. We also present a generalization of that model.

Although models *à la* Avellaneda-Stoikov have been built to deal with market making on stock markets, they can hardly be applied in practice to order-driven markets, because of the discrete nature of prices (linked to the tick size),[1] and because they do not take into account the very nature of the limit order books. These models are however well suited to quote-driven markets, such as the corporate bond market for instance, where dealers face issues very similar to those modeled in [13], or originally in [105]. For stock markets, a discrete modeling of the limit order book is required: we present a discrete model, inspired by the market making model of Guilbaud and Pham [92].

---

[1]In fact, the case of stocks with low tick sizes can be considered.

11.1   Introduction

What is a market maker? In short, a market maker is a liquidity provider: it is a trader[2] who, on a continuous and regular basis, proposes prices at which he stands ready to buy and sell a given asset. Market makers are met on almost all markets, and they play a central role in the price formation process. On order-driven markets, there are several kinds of market makers. Some are "official" market makers, in the sense that they have an agreement with an exchange for maintaining fair and orderly markets. This is the case of the Designated Market Makers (DMMs) on the NYSE.[3] In particular, DMMs must participate to the opening and closing auctions, and they must quote at the National Best Bid and Offer (NBBO) a specified percentage of the time. Some other market makers are just acting as liquidity providers without any obligation to do it: this is the case of some high-frequency traders who are almost continuously present on both sides of the limit order books (see [145]). On quote-driven markets, the market makers are the dealers: they provide liquidity to the other market participants by quoting bid and offer prices – see Chapter 2.

A market maker buys assets at the bid price he quotes, and sells assets at the ask (or offer) price he quotes. Therefore, a market maker expects to make profit out of the difference between his bid and offer prices, that is, his bid-ask spread.[4] However, a market maker seldom buys and sells simultaneously. Instead, he holds an inventory. For instance, at some time t_0, he may buy an asset at the bid price he quotes, and then hold it until someone comes to buy it, at time t_1. The latter transaction occurs at the ask price quoted by the market maker at time t_1. This ask price may be higher than the original bid price. In that case, the market maker makes a profit.[5] However, the ask price of the market maker at time t_1 may be lower than his bid price at time t_0, notably if the market price of the asset has dropped between t_0 and t_1. In that case, the market maker loses money.

The problem faced by a market maker is to propose/quote bid and offer prices in an optimal way for making money out of the bid-ask spread, while mitigating the risk associated with price changes, especially in the case of a large (long or short) inventory.

---

[2]Sometimes, the term "market maker" designates a company instead of a trader. Moreover, the trader is often replaced today by an algorithm.

[3]DMMs succeeded market specialists.

[4]In the case of a market organized around one or several limit order books, the bid-ask spread of a market maker should not be confused with the bid-ask spread of a specific venue, or the bid-ask spread of the market as a whole.

[5]The market maker can even make further profit on some venues offering rebates when liquidity-providing orders are executed.

The trade-off faced by a market maker is a complex one with both static and dynamic components.

If the market maker quotes a large spread, then each transaction leads to a large Mark-to-Market gain, but there are only very few such transactions. Conversely, if he quotes a narrow spread, then each transaction leads to a small Mark-to-Market gain, but there are many such transactions. This is the static part of the trade-off: high margin and low volume vs. low margin and high volume.

The problem faced by the market maker is also dynamic. If he constantly quotes conservatively (in particular with large spreads), then he may have difficulty to unwind his inventory, and this magnifies the risk associated with price changes. Conversely, if he quotes aggressively, then his P&L is naturally exposed to price risk, because the gain associated with the bid-ask spread, when the price does not change, is already small. Subsequently, in order to mitigate price risk, a market maker has dynamic (and skewed) quoting strategies. When he has a long inventory, he prices conservatively on the bid side (to avoid going on buying), and aggressively on the ask side (to increase the probability of selling assets, and therefore increase that of reducing his inventory). Conversely, a market maker with a short inventory prices aggressively on the bid side and conservatively on the ask side.

In this chapter, we present models aimed at finding a solution to the quoting problem faced by market makers. The reader may wonder why market making and optimal execution are dealt with in the same book. A first answer could be that we aim at being complete, by covering both the problems faced by liquidity takers and those faced by liquidity providers. A better answer is that there is almost no frontier between the academic literature on optimal execution and the academic literature on market making. The researchers working on these two topics are the same, and the models are often very similar – so much so that, for some authors, market making and optimal execution are part of the same research field: optimal trading.

In the case of stock markets, where everybody can be at the same time a liquidity provider and a liquidity taker, many market making models can very easily be transformed into optimal liquidation models with limit orders and market orders, and vice versa. The main difference between market making and optimal liquidation models is indeed that the former involve both buy and sell orders, while the latter involve either buy orders only or sell orders only. As a matter of fact, the first models of optimal liquidation with limit orders were inspired by the market making model of Avellaneda and Stoikov that we present in Section 11.2 – see especially the model of Guéant et al. [87].

The academic literature on market making is almost only devoted to market making on stock markets. This is related to the proximity with the literature on optimal execution, and to the interest of researchers for high-frequency trading on these markets. A significant proportion of high-frequency traders are indeed regarded as informal market makers, because their strategies consist in being active on both sides of the limit order books.[6] Inspired by an old paper on optimal dealer pricing written by Ho and Stoll [105] in the 1980s, and motivated by high-frequency trading applications, Avellaneda and Stoikov were the first to bring market making modeling back onto the table of the researchers in Quantitative Finance – see their 2008 paper [13].[7]

Since 2008, the literature has been burgeoning. Guéant et al. proved in [88] that the equations of the Avellaneda-Stoikov model could be simplified and transformed into a linear system of ordinary differential equations (ODEs). They also provided closed-form approximations for the optimal quotes. Guéant and Lehalle then generalized in [86] some of the results of [88] to a wider class of execution intensities (see Section 11.3). Cartea and Jaimungal, together with various coauthors, developed a wide variety of optimal trading models with both limit orders and market orders.[8] In a series of papers [37, 38, 45], they proposed several models, with various features, including market impact, short-term alpha, adverse selection, ambiguity aversion,[9] etc. Their modeling framework is different from the one proposed by Avellaneda and Stoikov, but they use a similar assumption: limit orders can be posted at any price.[10]

Almost all the academic papers on market making focus on stocks, and most of them – except a few, see below – are based on the assumption that limit orders can be posted at any price. However, this assumption is at odds with the reality of most stock markets because of the tick size. If we consider a stock with a bid-ask spread of 2 ticks, then a market maker providing liquidity on that stock has only a few choices for the bid (respectively ask) price he quotes: the best bid (respectively ask) price, the unique price inside the spread that is a multiple of the tick size – if he wants to quote aggressively – or a price corresponding to the second or third limit on the bid (respectively ask) side of the limit order book (or even beyond, but he has then little chance to provide liquidity in the short run) if he wishes to quote conservatively. Therefore, the very nature of the discrete problem faced by a market maker is not taken into account in most models. Furthermore, by enabling to post orders at any price, most of the models ignore the structure of limit order books, based on a queuing system, where priority matters. Although it is often convenient

---

[6]Unlike "official" market makers (DMMs, etc.), they have no constraint on the way they provide liquidity.

[7]Market making has always been a research topic for economists – see for instance [79]. However, the models proposed by the mathematicians are different. In particular, they are often dynamic models based on stochastic optimal control.

[8]They consider high-frequency market making strategies, and not pure liquidity provision strategies.

[9]See also the PhD dissertation [61] of one of their students.

[10]See [84] for a recent comparison between the two frameworks.

to build continuous models, it leads here to questionable models which can hardly be used on stock markets, except for small-tick stocks – i.e., when the average bid-ask spread is large with respect to the tick size.

Nevertheless, the models developed in the academic literature are far from being useless: they can often be applied, *mutatis mutandis*, for market making on quote-driven markets.[11] In particular, some dealers on bond markets are already using models adapted from academic models for determining the quotes they stream to electronic platforms.

In Section 11.2, we present the model initially proposed by Avellaneda-Stoikov in [13], and the method proposed in Guéant et al. [88] for solving it. In Section 11.3, we present a generalization of that model that can be adapted to most quote-driven markets.

We have previously insisted on the need for discrete models in the case of stocks, or, more generally, order-driven markets. A few papers of the literature present discrete models. In Cartea and Jaimungal [42] and Cartea et al. [43], the authors[12] proposed models for computing high-frequency trading strategies, in which traders can post on both sides of the limit order book, at the touch. Pham and his PhD students have also proposed several models for characterizing optimal high-frequency strategies on stock markets – see [70] and [92]. A model for market making on stock markets, inspired by the one of [92], is presented in Section 11.4.

11.2   The Avellaneda-Stoikov model

In this section, we present the market making model proposed by Avellaneda and Stoikov in [13].[13] This model was originally a model for market making on stocks. Even though it can hardly be used on stock markets, it constitutes a basis for other market making models, especially those addressing market making on quote-driven markets.

---

[11] In that case, limit orders are identified with quotes, and market orders are usually removed. Market orders could perhaps be used to proxy what happens on the IDB segment of the market.

[12] While I was finishing to write this book, Cartea, Jaimungal, and Penalva published a very interesting book entitled "Algorithmic and High-Frequency Trading" – see [44]. The approaches presented in their book constitute a very nice complement to those presented in this book, for both optimal liquidation and market making.

[13] In fact, we present the version of the Avellaneda-Stoikov model used in [88] and then in [86].

11.2.1   Framework

We consider a market maker operating on a single asset. The market price (or reference price) of this asset[14] is modeled by a process $(S_t)_t$ following a Brownian dynamics

$$dS_t = \sigma dW_t.$$

The market maker can continuously propose bid and ask prices to buy and sell the asset. We assume that every time a transaction occurs, one unit of the asset is traded.[15] Bid and ask quotes are modeled by two stochastic processes, respectively denoted by $(S_t^b)_t$ and $(S_t^a)_t$.

The times at which transactions occur are random. Furthermore, the distribution of the trade arrival times depends on the bid and ask prices quoted by the market maker.

We denote by $(N_t^b)_t$ and $(N_t^a)_t$ the two point processes modeling the number of assets that have been respectively bought and sold. The market maker's inventory is modeled by the process $(q_t)_t$. By definition, we have

$$q_t = N_t^b - N_t^a.$$

The intensity processes of $(N_t^b)_t$ and $(N_t^a)_t$ are denoted by $(\lambda_t^b)_t$ and $(\lambda_t^a)_t$. The main assumption of the Avellaneda-Stoikov model is that

$$\lambda_t^b = \Lambda^b(\delta_t^b) \quad \text{and} \quad \lambda_t^a = \Lambda^a(\delta_t^a), \tag{11.1}$$

where

$$\delta_t^b = S_t - S_t^b \quad \text{and} \quad \delta_t^a = S_t^a - S_t,$$

and where Λ^b and Λ^a are two positive and nonincreasing functions.

In other words, the instantaneous probability to trade at a given price is a function of the difference between the reference price and the price quoted by the market maker: the more conservatively a market maker quotes, the lower his chance to trade.

Avellaneda and Stoikov formally wrote the partial differential equations characterizing the optimal quotes for general functions Λ^b and Λ^a. Then, they focused on the specific case where

$$\Lambda^b(\delta) = \Lambda^a(\delta) = Ae^{-\kappa\delta}, \quad A > 0, \kappa > 0. \tag{11.2}$$

---

[14]In the original version of the Avellaneda-Stoikov model, this price was the mid-price of the stock. On a quote-driven market such as the corporate bond market, this price may be the Bloomberg composite price referred to as CBBT, or a price built from CBBT data and TRACE data (in the United States) – see Chapter 2.

[15]In [86], assets are traded Δ by Δ. This does not change anything, up to a rescaling of the model parameters. The only real assumption here is that the quantities traded are constant across trades.

In this setting, A characterizes the liquidity of the asset, and κ characterizes the price sensitivity of market participants.[16]

To be able to solve the problem in a rigorous way, we slightly modify the setting, and we assume, instead of Eq. (11.1), that

$$\lambda_t^b = \Lambda^b(\delta_t^b)1_{q_{t-}<Q} \quad \text{and} \quad \lambda_t^a = \Lambda^a(\delta_t^a)1_{q_{t-}>-Q}, \tag{11.3}$$

where $Q \in \mathbb{N}^*$ represents the maximum authorized inventory (long or short). In practice, this means that the market maker stops posting a bid (respectively ask) quote whenever $q_t = Q$ (respectively $q_t = -Q$).

The amount of cash on the market maker's account is modeled by the process $(X_t)_t$. By definition, it has the following dynamics:

$$dX_t = S_t^a dN_t^a - S_t^b dN_t^b = (S_t + \delta_t^a)dN_t^a - (S_t - \delta_t^b)dN_t^b.$$

For defining the optimization problem faced by the market maker, we consider that he has a time horizon T. The goal of the market maker is then to maximize the expected (CARA) utility criterion

$$\mathbb{E}\left[-\exp\left(-\gamma(X_T + q_T S_T - \ell(q_T))\right)\right], \tag{11.4}$$

over $(\delta_t^b)_t \in \mathcal{A}$ and $(\delta_t^a)_t \in \mathcal{A}$, where the set of admissible controls $\mathcal{A}$ is simply the set of predictable processes bounded from below.

As in the previous chapters, we apply a CARA utility function to the MtM value of the portfolio, penalized by a risk-liquidity premium $\ell(q_T)$ at the final time T. However, in the case of a market making problem, it is noteworthy that the risk-liquidity premium plays a minor role, because the inventory process is naturally mean-reverting around 0.

11.2.2 The Hamilton-Jacobi-Bellman equation and its solution

In order to maximize the objective function defined in Eq. (11.4) and to find the optimal controls, we use the classical tools of stochastic optimal control theory. However, we do not need here to use the concept of viscosity solution, because it turns out that the value function can be computed (almost) in closed-form.

---

[16] In this model, we consider a market maker in isolation. Competition between market makers is not modeled, but A and κ can be impacted by the degree of competition between market makers on the asset. See [69] for estimations of hit ratios in the case of market makers in competition on bond markets.

The Hamilton-Jacobi-Bellman (HJB) equation associated with our problem is given by

$$0 = \partial_t u(t,x,q,S) + \frac{1}{2}\sigma^2 \partial_{SS}^2 u(t,x,q,S) \tag{11.5}$$

$$+1_{q<Q}\sup_{\delta^b} \Lambda^b(\delta^b)\left[u(t,x-S+\delta^b,q+1,S) - u(t,x,q,S)\right]$$

$$+1_{q>-Q}\sup_{\delta^a} \Lambda^a(\delta^a)\left[u(t,x+S+\delta^a,q-1,S) - u(t,x,q,S)\right],$$

for $q \in \{-Q,\ldots,Q\}$, and $(t,S,x) \in [0,T] \times \mathbb{R}^2$, with the terminal condition

$$u(T,x,q,S) = -\exp\left(-\gamma(x+qS-\ell(q))\right). \tag{11.6}$$

In the case of exponential intensities (Eq. (11.2)), Avellaneda and Stoikov showed in [13] that $e^{-\gamma x}$ could be factored out of u in the above equations, hence reducing the dimensionality of the problem from 4 to 3. In [88], Guéant et al. improved this result by proving that a solution u to Eqs. (11.5) and (11.6) could be found by solving a linear system of ODEs. Subsequently, they reduced the dimensionality of the problem even further: from 3 to 2. In the next paragraphs, we present the results obtained in [88].

For solving Eqs. (11.5) and (11.6) in the case of exponential intensities, the key idea is to consider the following ansatz:

$$u(t,x,q,S) = -\exp(-\gamma(x+qS))v_q(t)^{-\frac{\gamma}{\kappa}}. \tag{11.7}$$

This ansatz is complex, as it is based on three ideas. When a CARA utility function is considered, it is always relevant to factor out the MtM value of the market maker's portfolio $(x+qs)$, and therefore to write the (candidate) value function as $u(t,x,q,S) = -\exp(-\gamma(x+qS+\theta(t,q,S)))$. This first idea was used in Chapters 9 and 10 to reduce the dimensionality of the problems under consideration.[17] The second idea is that θ has no reason to depend on S. The third – and most original – idea is to consider $(t,q) \mapsto v_q(t) = e^{\kappa\theta(t,q)}$ instead of θ. The good thing is that Eqs. (11.5) and (11.6) then boil down to a linear system of ODEs and a simple terminal condition.

It is indeed straightforward to see that if $(v_q)_{|q|\leq Q}$ are positive functions satisfying

$$\forall q \in \{-Q,\ldots,Q\}, \forall t \in [0,T],$$

$$\frac{d}{dt}v_q(t) = \frac{\kappa}{2}\gamma\sigma^2 q^2 v_q(t) - A\left(1+\frac{\gamma}{\kappa}\right)^{-\left(1+\frac{\kappa}{\gamma}\right)}\left(1_{q>-Q}v_{q-1}(t) + 1_{q<Q}v_{q+1}(t)\right), \tag{11.8}$$

---

[17]We can consider θ or $-\theta$.

with the terminal condition

$$\forall q \in \{-Q, \ldots, Q\}, v_q(T) = e^{-\kappa\ell(q)},$$

then the function u defined by Eq. (11.7) is a solution of Eqs. (11.5) and (11.6).

Eq. (11.8) is a linear system of ODEs. It can therefore be solved very easily.

If we denote by M the matrix

$$M = \begin{pmatrix} \alpha Q^2 & -\eta & 0 & \cdots & \cdots & & \cdots & 0 \\ -\eta & \alpha(Q-1)^2 & -\eta & 0 & \ddots & & \ddots & \vdots \\ 0 & \ddots & \ddots & \ddots & \ddots & \ddots & & \vdots \\ \vdots & \ddots & & \ddots & \ddots & \ddots & \ddots & \vdots \\ \vdots & \ddots & & \ddots & \ddots & \ddots & \ddots & \vdots \\ \vdots & \ddots & & \ddots & \ddots & \ddots & \ddots & 0 \\ \vdots & \ddots & & \ddots & 0 & -\eta & \alpha(Q-1)^2 & -\eta \\ 0 & \cdots & & \cdots & \cdots & 0 & -\eta & \alpha Q^2 \end{pmatrix},$$

where $\alpha = \frac{\kappa}{2}\gamma\sigma^2$ and $\eta = A\left(1 + \frac{\gamma}{\kappa}\right)^{-(1+\frac{\kappa}{\gamma})}$, then

$$(v_{-Q}(t), \ldots, v_Q(t)) = \left(e^{-\kappa\ell(-Q)}, \ldots, e^{-\kappa\ell(Q)}\right) \exp(-M(T-t)) \qquad (11.9)$$

defines the unique solution $(v_q)_{|q|\leq Q}$ of Eq. (11.8) with the appropriate terminal condition.

By using classical techniques of mathematical analysis and stochastic optimal control (the maximum principle, verification arguments, etc.), Guéant et al. proved in [88] that the functions $(v_q)_{|q|\leq Q}$ defined by Eq. (11.9) are positive functions, and that the resulting function u defined by Eq. (11.7) is not only solution of the HJB equation – Eqs. (11.5) and (11.6) – but in fact the value function associated with the optimization problem. Furthermore, they proved that the optimal quotes of the market maker are given by the following theorem:

Theorem 11.1. *The optimal controls* $(\delta_t^{b*})_t$ *and* $(\delta_t^{a*})_t$ *are given by*

$$\delta_t^{b*} = \delta^{b*}(t, q_t) \quad and \quad \delta_t^{a*} = \delta^{a*}(t, q_t),$$

where

$$\delta^{b*}(t, q) = \frac{1}{\kappa} \ln\left(\frac{v_q(t)}{v_{q+1}(t)}\right) + \frac{1}{\gamma} \ln\left(1 + \frac{\gamma}{\kappa}\right), \quad q < Q, \qquad (11.10)$$

$$\delta^{a*}(t, q) = \frac{1}{\kappa} \ln\left(\frac{v_q(t)}{v_{q-1}(t)}\right) + \frac{1}{\gamma} \ln\left(1 + \frac{\gamma}{\kappa}\right), \quad q > -Q. \qquad (11.11)$$

In particular, Theorem 11.1 states that the optimal bid and ask quotes of the market maker in the Avellaneda-Stoikov model are given by the following functions:

$$S^{b*}(t, S, q) = S - \frac{1}{\kappa} \ln \left(\frac{v_q(t)}{v_{q+1}(t)} \right) - \frac{1}{\gamma} \ln \left(1 + \frac{\gamma}{\kappa} \right), \qquad (11.12)$$

$$S^{a*}(t, S, q) = S + \frac{1}{\kappa} \ln \left(\frac{v_q(t)}{v_{q-1}(t)} \right) + \frac{1}{\gamma} \ln \left(1 + \frac{\gamma}{\kappa} \right). \qquad (11.13)$$

These optimal quotes are made of three parts: (i) the current value of reference price S, (ii) the term $\frac{1}{\gamma} \ln \left(1 + \frac{\gamma}{\kappa} \right)$, which corresponds to the static part of the trade-off faced by the market maker, and (iii) the terms involving the functions $(v_q)_{|q| \leq Q}$, which make the market maker's optimal quotes depend on its inventory.

In practice, $(v_q)_{|q| \leq Q}$ can be computed numerically very easily, either by using the closed-form expression of Eq. (11.9), or by numerically approximating – with a Euler scheme, for instance – the unique solution of the linear system of ODEs (11.8) that satisfies the terminal condition $v_q(T) = e^{-\kappa \ell(q)}$. Subsequently, the optimal bid and ask quotes of a market maker in the Avellaneda-Stoikov can be computed very easily.

11.2.3 The Guéant–Lehalle–Fernandez-Tapia formulas

In Eqs. (11.12) and (11.13), the optimal bid and ask quotes of the market maker depend on both the current market price of the asset and the current inventory. They also depend on the time variable t. For a market maker on stock markets, this dependence on t may seem natural, because positions are often unwound before the end of the day to avoid overnight risk. However, for a market maker on bonds, this may seem rather bizarre. In fact, the optimal bid and ask quotes derived in Eqs. (11.12) and (11.13) only depend on time because we have – somehow artificially – introduced a final time T.

Guéant et al. studied in [88] the asymptotic behavior of the optimal bid and ask quotes, when $T \to +\infty$. Their result is given in the following theorem:

Theorem 11.2. *The functions defining the optimal bid and ask quotes in Theorem 11.1 have an asymptotic behavior when $T \to +\infty$ given by*

$$\lim_{T \to \infty} \delta^{b*}(0, q) = \delta_\infty^{b*}(q) = \frac{1}{\gamma} \ln \left(1 + \frac{\gamma}{\kappa} \right) + \frac{1}{\kappa} \ln \left(\frac{f_q^0}{f_{q+1}^0} \right), \qquad (11.14)$$

and

$$\lim_{T \to \infty} \delta^{a*}(0, q) = \delta_\infty^{a*}(q) = \frac{1}{\gamma} \ln \left(1 + \frac{\gamma}{\kappa} \right) + \frac{1}{\kappa} \ln \left(\frac{f_q^0}{f_{q-1}^0} \right), \qquad (11.15)$$

where f^0 is uniquely characterized (up to a multiplicative constant) by

$$f^0 \in \operatorname*{argmin}_{f \in \mathbb{R}^{2Q+1} \setminus \{0\}} \frac{\sum_{q=-Q}^{Q} \alpha q^2 f_q^2 + \eta \sum_{q=-Q}^{Q-1}(f_{q+1} - f_q)^2 + \eta f_Q^2 + \eta f_{-Q}^2}{\sum_{q=-Q}^{Q} f_q^2}.$$

$$(11.16)$$

Proof. *(Sketch of the proof)*[18]

Let $v(t) = (v_{-Q}(t), \ldots, v_Q(t))'$ be the transpose of $(v_{-Q}(t), \ldots, v_Q(t))$. With this definition, Eq. (11.9) writes $v(t) = \exp(-M(T-t))v(T)$.

M is symmetric matrix. Let $\lambda^0 \leq \ldots \leq \lambda^{2Q}$ be its eigenvalues, and $(f^0, \ldots, f^{2Q})$ an orthonormal basis of eigenvectors associated with these eigenvalues. We have

$$v(0) = \sum_{i=0}^{2Q} \exp(-\lambda^i T)(f^i \cdot v(T))f^i.$$

The technical point of the proof consists in proving that the smallest eigenvalue of M – here λ^0 – is simple, and that the corresponding eigenvector f^0 can be chosen with all its entries positive. This is done in [88] by considering the matrix $M + 2\eta I_{2Q+1}$, and by showing that its inverse, i.e., $(M + 2\eta I_{2Q+1})^{-1}$, verifies the property of Perron-Frobenius.[19]

Then, $\forall q \in \{-Q, \ldots, Q\}$,

$$v_q(0) \sim_{T \to +\infty} \exp(-\lambda^0 T) \underbrace{(f^0 \cdot v(T))}_{>0} \underbrace{f_q^0}_{>0}.$$

By using Eqs. (11.10) and (11.11), we have

$$\lim_{T \to \infty} \delta^{b*}(0, q) = \frac{1}{\gamma} \ln\left(1 + \frac{\gamma}{\kappa}\right) + \frac{1}{\kappa} \ln\left(\frac{f_q^0}{f_{q+1}^0}\right),$$

$$\lim_{T \to \infty} \delta^{a*}(0, q) = \frac{1}{\gamma} \ln\left(1 + \frac{\gamma}{\kappa}\right) + \frac{1}{\kappa} \ln\left(\frac{f_q^0}{f_{q-1}^0}\right).$$

Because f^0 is an eigenvector corresponding to the smallest eigenvalue of M, it is also an eigenvector corresponding to the smallest eigenvalue of the matrix $M + 2\eta I_{2Q+1}$. By using Courant-Fischer theorem, it is therefore characterized (up to a multiplicative constant) as a minimizer of the Rayleigh quotient

$$\frac{f \cdot (M + 2\eta I_{2Q+1})f}{f \cdot f}.$$

---

[18] For a complete proof, see [88].
[19] This is true for the inverse of any matrix which is the sum of a nonnegative diagonal matrix and the matrix of a discrete Laplace operator.

In other words,

$$f^0 \in \underset{f \in \mathbb{R}^{2Q+1}\setminus\{0\}}{\mathrm{argmin}} \quad \frac{\sum_{q=-Q}^{Q} \alpha q^2 f_q^2 + \eta \sum_{q=-Q}^{Q-1} (f_{q+1} - f_q)^2 + \eta f_Q^2 + \eta f_{-Q}^2}{\sum_{q=-Q}^{Q} f_q^2}.$$

The formulas obtained in Theorem 11.2 are the relevant ones for the optimal bid and ask quotes of a market maker with no time horizon. These formulas can also be used to approximate the optimal bid and ask quotes of a market maker, as soon as the time of liquidation of the portfolio is not close. In particular, these formulas do not depend on the penalty function $\ell(\cdot)$.

Guéant et al. introduced an additional idea in [88] for approximating in closed-form the optimal bid and ask quotes of Theorem 11.2.[20] It consists in replacing the vector $(f_{-Q}^0, \ldots, f_Q^0)$ by a vector of the form $(\tilde{f}^0(-Q), \ldots, \tilde{f}^0(Q))$, where $\tilde{f}^0$ is a function of $L^2(\mathbb{R})$ characterized by[21]

$$\tilde{f}^0 \in \underset{\tilde{f} \in L^2(\mathbb{R})}{\mathrm{argmin}} \frac{\int_{-\infty}^{\infty} \left(\alpha x^2 \tilde{f}(x)^2 + \eta \tilde{f}'(x)^2 \right) dx}{\int_{-\infty}^{\infty} \tilde{f}(x)^2 dx},$$

i.e., a continuous counterpart of the characterization (11.16).

Finding $\tilde{f}^0$ is just another problem of variational calculus. It is straightforward to check that

$$\tilde{f}^0(x) \propto \exp\left(-\frac{1}{2}\sqrt{\frac{\alpha}{\eta}} x^2 \right).$$

By replacing $(f_q^0)_q$ by $(\tilde{f}^0(q))_q$ in Eqs. (11.14) and (11.15), we obtain the following approximations, known as the Guéant–Lehalle–Fernandez-Tapia formulas:[22]

$$\delta_\infty^{b*}(q) \simeq \frac{1}{\gamma} \ln\left(1 + \frac{\gamma}{\kappa}\right) + \frac{2q+1}{2}\sqrt{\frac{\sigma^2 \gamma}{2\kappa A}\left(1 + \frac{\gamma}{\kappa}\right)^{1+\frac{\kappa}{\gamma}}}, \quad (11.17)$$

$$\delta_\infty^{a*}(q) \simeq \frac{1}{\gamma} \ln\left(1 + \frac{\gamma}{\kappa}\right) - \frac{2q-1}{2}\sqrt{\frac{\sigma^2 \gamma}{2\kappa A}\left(1 + \frac{\gamma}{\kappa}\right)^{1+\frac{\kappa}{\gamma}}}. \quad (11.18)$$

---

[20] Avellaneda and Stoikov also proposed approximations of the optimal quotes in [13]. However, their approximations correspond to the case $T \to 0$.

[21] $\tilde{f}^0$ is in fact in the set of $L^2(\mathbb{R})$ functions satisfying

$$\int_{-\infty}^{\infty} \left(\alpha x^2 \tilde{f}(x)^2 + \eta \tilde{f}'(x)^2 \right) dx < +\infty,$$

where the first derivative has to be understood in the sense of a weak derivative.

[22] These formulas have recently been generalized in [84] to the case of more general functions Λ^b and Λ^a.

We see from Eqs. (11.17) and (11.18) that the optimal bid-ask spread of the market maker is approximated by

$$\psi_\infty^*(q) = \delta_\infty^{b*}(q) + \delta_\infty^{a*}(q) \simeq \frac{2}{\gamma} \ln\left(1 + \frac{\gamma}{\kappa}\right) + \sqrt{\frac{\sigma^2\gamma}{2\kappa A}\left(1 + \frac{\gamma}{\kappa}\right)^{1+\frac{\kappa}{\gamma}}}. \qquad (11.19)$$

In particular, it does not depend on the inventory q. This property only holds for the above approximations, but it indicates that the optimal strategy of the market maker mainly consists in moving in parallel the bid and ask quotes as a function of the inventory. Furthermore, Eqs. (11.17) and (11.18) suggest that the dependence of the bid and ask quotes on the inventory is approximately affine.

The skewness of the market strategy, defined as the difference between δ^b and δ^a, is approximated by

$$\delta_\infty^{b*}(q) - \delta_\infty^{a*}(q) \simeq 2q\sqrt{\frac{\sigma^2\gamma}{2\kappa A}\left(1 + \frac{\gamma}{\kappa}\right)^{1+\frac{\kappa}{\gamma}}}. \qquad (11.20)$$

It measures the market maker's propensity to quote aggressively on the bid (respectively ask) side and conservatively on the ask (respectively bid) side when he has a short (respectively long) position. We see that it depends linearly on the inventory.

The above approximations help to clarify the role of each parameter and the effects at play.

A market maker faces two risks of different natures. First, there is a static risk, associated with transaction uncertainty.[23] Even if the reference price was constant (i.e., if the volatility parameter was equal to 0), the bid-ask spread of the market maker would depend on the risk aversion parameter γ. In the no-volatility case, the optimal bid-ask spread is indeed

$$\psi^* = \frac{2}{\gamma} \ln\left(1 + \frac{\gamma}{\kappa}\right).$$

It balances infrequent trades and high margin per trade vs. frequent trades and low margin per trade. In particular, when $\sigma = 0$, the higher γ, the smaller the bid-ask spread, in order to reduce uncertainty.

The second kind of risk is related to price risk. However, it is more subtle than the classical risk that the price moves. The second risk faced by the market maker is actually the risk that the price moves adversely without him being able to unwind his position rapidly enough (because of trade uncertainty). The higher the risk aversion to this combination of price risk and

---

[23] In the Avellaneda-Stoikov model, a market maker cannot be a liquidity taker: he can only influence his chance to buy and sell by adjusting his quotes.

transaction uncertainty, the larger the bid-ask spread and the skewness of the strategy, because the market maker wants to avoid holding large inventories.

The previous analysis is confirmed by Eqs. (11.19) and (11.20). Because γ measures risk aversion to both kinds of risk, the overall effect of risk aversion on the spread is ambiguous: the first term of Eq. (11.19) is decreasing in γ, while the second is increasing in γ. As far as the skewness is concerned, Eq. (11.20) states that the propensity to adjust quotes is an increasing function of γ.

Regarding the volatility of the asset price, we see in Eq. (11.19) that the higher σ, the larger the bid-ask spread: when the risk associated with price changes increases, the market maker widens his bid-ask spread. This is a natural reaction to an increase in the second kind of risk described above. Similarly, we see in Eq. (11.20) that the skewness of the optimal market making strategy is (in absolute value) an increasing – in fact, linear – function of σ: the higher σ, the higher the magnitude of quote changes in order to stay close to a flat portfolio.

As far as the parameters of the functions Λ^b and Λ^a are concerned, we see in Eqs. (11.19) and (11.20) that the larger A, the smaller the bid-ask spread and the skewness (in absolute value) of the strategy. This is because A matters only for the combination of price risk and transaction uncertainty described previously. The relevant variable for this complex risk is in fact $\frac{\sigma^2}{A}$: A and σ have therefore opposite effects. Coming to κ, there are several effects at stake: (i) an increase in κ has the same effect as a decrease in A, and (ii) an increase in κ decreases the probability to trade at prices that are far from the reference price – when $\delta^{a,b} > 0$. Eq. (11.19) suggests that the latter effect dominates: the larger κ, the smaller the bid-ask spread. Eq. (11.20) also suggests that the skewness (in absolute value) is a decreasing function of κ. However, one may be very careful when interpreting the effect of κ because of what happens when $\delta^b < 0$ or $\delta^a < 0$.[24]

11.3   Generalization of the Avellaneda-Stoikov model

11.3.1   Introduction

The model presented in the previous section can be generalized in several ways. A group of researchers driven by Cartea and Jaimungal proposed many optimal trading models, with various features, including a drift in

---

[24]The exponential forms chosen for Λ^b and Λ^a raise indeed a lot of concerns.

the price process, a short-term α, some form of adverse selection, etc. – see [37, 38, 42, 43, 44, 45]. Some of these features can be added to enrich the original Avellaneda-Stoikov model, without much change in the mathematics – see [88].

We have seen previously that market making models should be discrete to be suited for the stock market. This is clearly not the case of the Avellaneda-Stoikov. Nevertheless, the modeling framework introduced by Avellaneda and Stoikov can be used for market making on many quote-driven markets: the problem of the tick size is often irrelevant on these markets, and there is no queuing system, because there is no limit order book. However, to be used in practice, a model à la Avellaneda-Stoikov needs to have at least two important features. First, the intensity functions Λ^b and Λ^a must not be limited to the exponential functions of Eq. (11.2): a large class of functions must be allowed. Second, a realistic model must allow for the market making of several assets at the same time.

The exponential forms chosen for Λ^b and Λ^a are interesting because the problem then boils down to solving a linear system of ordinary differential equations. However, on quote-driven markets, intensity functions[25] are sometimes better represented (around the reference price[26]) by logistic functions of the form

$$\frac{Ae^{-\kappa\delta}}{1 + Be^{-\kappa\delta}}, \quad A, B, \kappa > 0.$$

In the following paragraphs, we present a general model in which a large class of intensity functions is allowed.

In the Avellaneda-Stoikov model, the market maker buys and sells a given asset, but, in practice, market makers are often in charge of a set of several assets. Is it relevant to consider these different assets in isolation? When asset price changes are uncorrelated, the answer is yes. However, when asset price changes are correlated, positively or negatively, a multi-asset market making model is needed.

Let us consider the example of corporate bonds. There are usually dozens of bonds issued by the same company. Therefore, optimal quotes for a specific bond should not depend on the market maker's inventory in that bond, but

---

[25]Practitioners usually compute hit ratios, which represent the probability to trade, following a request for quote (RFQ), given that the RFQ has led to a trade with one of the dealers in competition. For designing market making models, hit ratios are not sufficient: one needs to consider both the probability to receive an RFQ, and the probability that the client accepts to trade at the quoted price.

[26]When δ takes very negative values, the probability to trade should be very high. However, it is not relevant to allow for extreme values of δ in models. In particular, logistic functions prevent choosing of extreme values for δ.

instead on the risk profile of the whole bond portfolio with respect to the issuer. In particular, when a market maker has a long inventory in a bond and an almost equivalent short inventory in another bond of the same issuer, there may be no reason for him to skew his bid and ask quotes on these two bonds, contrary to what single-asset market making models would suggest. In the following section, we present a general multi-asset market making model.

11.3.2 A general multi-asset market making model

11.3.2.1 Framework

We consider a market maker operating on d assets. The reference price of asset i is modeled by a process $(S_t^i)_t$ following a Brownian dynamics

$$dS_t^i = \sigma^i dW_t^i.$$

We assume that the d-dimensional process $(\sigma^1 W_t^1, \ldots, \sigma^d W_t^d)_t$ has a nonsingular covariance matrix $\Sigma = (\rho^{i,j}\sigma^i\sigma^j)_{1 \leq i,j \leq d}$.

The market maker proposes bid and ask prices to buy and sell each of these d assets. As in the single-asset case, we assume that each time a transaction involving asset i occurs, one unit of this asset is traded. The bid and ask quotes of the market maker are modeled by $2d$ stochastic processes, respectively denoted by $(S_t^{i,b})_t$ and $(S_t^{i,a})_t$, $i \in \{1, \ldots, d\}$.

We denote by $(N_t^{i,b})_t$ and $(N_t^{i,a})_t$ the point processes modeling the volumes of asset i that have been respectively bought and sold. The market maker's inventory is modeled by the d-dimensional process $(q_t^1, \ldots, q_t^d)_t$. By definition, we have

$$\forall i, q_t^i = N_t^{i,b} - N_t^{i,a}.$$

The intensity processes of $(N_t^{i,b})_t$ and $(N_t^{i,a})_t$ are denoted by $(\lambda_t^{i,b})_t$ and $(\lambda_t^{i,a})_t$. As in the previous model, we assume that

$$\lambda_t^{i,b} = \Lambda^{i,b}(\delta_t^{i,b})1_{q_{t_-}^i < Q^i} \quad \text{and} \quad \lambda_t^{i,a} = \Lambda^{i,a}(\delta_t^{i,a})1_{q_{t_-}^i > -Q^i}, \tag{11.21}$$

where

$$\delta_t^{i,b} = S_t^i - S_t^{i,b} \quad \text{and} \quad \delta_t^{i,a} = S_t^{i,a} - S_t^i,$$

where $(Q^i)_i$ are inventory bounds on each of the d assets, and where $\forall i, \Lambda^{i,b}$ and $\Lambda^{i,a}$ are two positive and nonincreasing functions.

The amount of cash on the market maker's account is modeled by the process $(X_t)_t$.

By definition, $(X_t)_t$ has the following dynamics:

$$
\begin{aligned}
dX_t &= \sum_{i=1}^{d} S_t^{i,a} dN_t^{i,a} - S_t^{i,b} dN_t^{i,b} \\
&= \sum_{i=1}^{d} (S_t^i + \delta_t^{i,a}) dN_t^{i,a} - (S_t^i - \delta_t^{i,b}) dN_t^{i,b}.
\end{aligned}
$$

As in the single-asset model, we consider a time horizon T. The goal of the market maker is to maximize the expected (CARA) utility criterion

$$
\mathbb{E}\left[-\exp\left(-\gamma\left(X_T + \sum_{i=1}^{d} q_T^i S_T^i\right)\right)\right], \tag{11.22}
$$

over $(\delta_t^{i,b})_t \in \mathcal{A}^d$ and $(\delta_t^{i,a})_t \in \mathcal{A}^d$, where the set of admissible controls $\mathcal{A}$ is the same as in Section 11.2.[27]

11.3.2.2 Computing the optimal quotes

The Hamilton-Jacobi-Bellman (HJB) equation associated with the problem of maximizing the objective function defined by Eq. (11.22) is given by

$$
0 = \partial_t u(t, x, q, S) + \frac{1}{2} \sum_{1 \leq i,j \leq d} \rho^{i,j} \sigma^i \sigma^j \partial_{S^i S^j}^2 u(t, x, q, S) \tag{11.23}
$$

$$
+ \sum_{1 \leq i \leq d} 1_{q^i < Q^i} \sup_{\delta^{i,b}} \Lambda^{i,b}(\delta^{i,b}) \left[u(t, x - S^i + \delta^{i,b}, q + e^i, S) - u(t, x, q, S) \right]
$$

$$
+ \sum_{1 \leq i \leq d} 1_{q^i > -Q^i} \sup_{\delta^{i,a}} \Lambda^{i,a}(\delta^{i,a}) \left[u(t, x + S^i + \delta^{i,a}, q - e^i, S) - u(t, x, q, S) \right],
$$

for $q^i \in \{-Q^i, \ldots, Q^i\}, \forall i$, and $(t, S, x) \in [0,T] \times \mathbb{R}^{d+1}$,[28] with the terminal condition

$$
u(T, x, q, S) = -\exp\left(-\gamma\left(x + \sum_{i=1}^{d} q^i S^i\right)\right). \tag{11.24}
$$

When the intensity functions $(\Lambda^{i,b})_i$ and $(\Lambda^{i,a})_i$ are not of exponential form, Eq. (11.23) cannot be transformed into a linear system of ODEs. However, by using the classical change of variables

$$
u(t, x, q, S) = -\exp\left(-\gamma\left(x + \sum_{1 \leq i \leq d} q^i S^i + \theta(t, q)\right)\right), \tag{11.25}
$$

the dimensionality of the problem can still be reduced from $2 + 2d$ to $1 + d$.

---

[27] For the sake of simplicity, we omit the penalty term.
[28] e^i is the ith vector of the canonical basis of $\mathbb{R}^d$.

It is indeed straightforward to see that if the function θ satisfies the following equations:

$\forall i, \forall q^i \in \{-Q^i, \ldots, Q^i\}, \forall t \in [0, T]$,

$$0 = \partial_t \theta(t, q) - \frac{1}{2} \gamma q' \Sigma q \tag{11.26}$$

$$+ \sum_{1 \leq i \leq d} 1_{q^i < Q^i} \sup_{\delta^{i,b}} \frac{\Lambda^{i,b}(\delta^{i,b})}{\gamma} \left(1 - \exp\left(-\gamma \left(\delta^{i,b} + \theta(t, q + e^i) - \theta(t, q)\right)\right)\right)$$

$$+ \sum_{1 \leq i \leq d} 1_{q^i > -Q^i} \sup_{\delta^{i,a}} \frac{\Lambda^{i,a}(\delta^{i,a})}{\gamma} \left(1 - \exp\left(-\gamma \left(\delta^{i,a} + \theta(t, q - e^i) - \theta(t, q)\right)\right)\right),$$

with the terminal condition

$$\forall i, \forall q^i \in \{-Q^i, \ldots, Q^i\}, \theta(T, q) = 0,$$

then the function u defined by Eq. (11.25) is a solution of Eqs. (11.23) and (11.24).

For the suprema in Eq. (11.26) to be well defined, with a unique maximizer, simple assumptions are[29]

- $\forall i, \Lambda^{i,b}$ and $\Lambda^{i,a}$ are C^2 decreasing functions with

$$\lim_{\delta \to +\infty} \Lambda^{i,b}(\delta) = \lim_{\delta \to +\infty} \Lambda^{i,a}(\delta) = 0.$$

- The intensity functions $(\Lambda^{i,b})_i$ and $(\Lambda^{i,a})_i$ satisfy

$$\Lambda^{i,b}(\delta) \Lambda^{i,b\prime\prime}(\delta) \leq 2 \left(\Lambda^{i,b\prime}(\delta)\right)^2 \quad \text{and} \quad \Lambda^{i,a}(\delta) \Lambda^{i,a\prime\prime}(\delta) \leq 2 \left(\Lambda^{i,a\prime}(\delta)\right)^2.$$

In that case – see [86] for an analogy – we can prove the following theorem:

Theorem 11.3. *There exists a unique solution θ, C^1 in time, of Eq. (11.26) with the terminal condition*

$$\forall i, \forall q^i \in \{-Q^i, \ldots, Q^i\}, \theta(T, q) = 0.$$

The function u defined by Eq. (11.25) is the value function of the optimal control problem.

---

[29] Other assumptions are possible. The uniqueness property of the maximizer is not required in general.

Moreover, the optimal bid and ask quotes are characterized by

$$\delta_t^{i,b*} = \delta^{i,b*}\left(\theta(t,q_t) - \theta(t,q_t + e^i)\right),$$

and

$$\delta_t^{i,a*} = \delta^{i,a*}\left(\theta(t,q_t) - \theta(t,q_t - e^i)\right),$$

where the functions $p \mapsto \delta^{i,b*}(p)$ *and* $p \mapsto \delta^{i,a*}(p)$ *are implicitly[30] defined by*

$$p = \delta^{i,b*}(p) - \frac{1}{\gamma}\ln\left(1 - \gamma\frac{\Lambda^{i,b}\left(\delta^{i,b*}(p)\right)}{\Lambda^{i,b\prime}\left(\delta^{i,b*}(p)\right)}\right),$$

and

$$p = \delta^{i,a*}(p) - \frac{1}{\gamma}\ln\left(1 - \gamma\frac{\Lambda^{i,a}\left(\delta^{i,a*}(p)\right)}{\Lambda^{i,a\prime}\left(\delta^{i,a*}(p)\right)}\right).$$

Proof. *(Sketch of the proof)[31]*

By using our assumptions on the intensity functions, it is easy to prove that, for all $p \in \mathbb{R}$, *the function*

$$\delta \mapsto \frac{\Lambda^{i,b}(\delta)}{\gamma}\left(1 - \exp\left(-\gamma\left(\delta - p\right)\right)\right)$$

has a unique maximizer $\delta^{i,b*}(p)$ *characterized by the first-order condition*

$$p = \delta^{i,b*}(p) - \frac{1}{\gamma}\ln\left(1 - \gamma\frac{\Lambda^{i,b}\left(\delta^{i,b*}(p)\right)}{\Lambda^{i,b\prime}\left(\delta^{i,b*}(p)\right)}\right).$$

Furthermore, the function $p \mapsto \delta^{i,b*}(p)$ *is a* C^1 *function.*

The same is true on the ask side: for all $p \in \mathbb{R}$, *the function*

$$\delta \mapsto \frac{\Lambda^{i,a}(\delta)}{\gamma}\left(1 - \exp\left(-\gamma\left(\delta - p\right)\right)\right)$$

has a unique maximizer $\delta^{i,a*}(p)$ *characterized by the first-order condition*

$$p = \delta^{i,a*}(p) - \frac{1}{\gamma}\ln\left(1 - \gamma\frac{\Lambda^{i,a}\left(\delta^{i,a*}(p)\right)}{\Lambda^{i,a\prime}\left(\delta^{i,a*}(p)\right)}\right).$$

Furthermore, the function $p \mapsto \delta^{i,a*}(p)$ *is a* C^1 *function.*

---

[30]The assumptions made on the intensity functions guarantee that these functions are well defined.

[31]The complete proof is long. See [86] for an analogous proof in the case of an optimal execution strategy.

As a consequence, we can apply Cauchy-Lipschitz to Eq. (11.26) to prove that a (unique) local C^1 solution exists. For proving that a global solution exists on $[0,T]$, a first point is to notice that $t \mapsto \theta(t,q) + \frac{1}{2}\gamma q'\Sigma q(T-t)$ is a decreasing function. Therefore, we only need to find an upper bound to θ in order to prove global existence.

For proving that θ does not blow up, we first notice that

$$\bar{\theta}(t,q) = \bar{\theta}(t) = \sum_{1 \leq i \leq d} \sup_\delta \frac{\Lambda^{i,b}(\delta)}{\gamma} (1 - \exp(-\gamma\delta))(T-t)$$

$$+ \sum_{1 \leq i \leq d} \sup_\delta \frac{\Lambda^{i,a}(\delta)}{\gamma} (1 - \exp(-\gamma\delta))(T-t)$$

is a supersolution to Eq. (11.26). Then, the maximum principle guarantees that $\theta(t,q) \leq \bar{\theta}(t,q)$.

Now, by using straightforward computations and a classical verification argument, the function u defined by Eq. (11.25) is indeed the value function associated with the optimization problem, and the optimal quotes are given by

$$\delta_t^{i,b*} = \delta^{i,b*}\left(\theta(t,q_t) - \theta(t,q_t + e^i)\right),$$

and

$$\delta_t^{i,a*} = \delta^{i,a*}\left(\theta(t,q_t) - \theta(t,q_t - e^i)\right).$$

Theorem 11.3 states that the optimal bid and ask quotes of a market maker can be numerically approximated in a very simple way. First, the functions $(\delta^{i,b*})_i$ and $(\delta^{i,a*})_i$ have to be tabulated. For this purpose, many methods are available: a Newton's method, the bisection method, etc. Then, the optimal quotes can be numerically approximated by approximating the solution of the large system of nonlinear ODEs (11.26). For this purpose, a simple Euler scheme does the job.

It is noteworthy that, far from T, the optimal quotes are almost independent from the time variable t; in practice, the solution of the system (11.26) needs to be approximated backward in time until the resulting approximations of the optimal quotes almost stop evolving.

The model we have presented is a generalization of the Avellaneda-Stoikov model that can be used by market practitioners on many quote-driven markets, such as the corporate bond market. This model can be further generalized to feature additional effects, such as adverse selection, a drift in the price process, etc. However, one important point is omitted in the model: the possibility for a dealer to trade on the IDB segment, or on platforms where market orders can be posted – although, on most of the markets that are mainly quote-driven, they represent a tiny percentage of the total volume traded. In practice dealers are not always providing liquidity: in some specific situations, they can also reduce their exposure without waiting for an RFQ from a client.

11.4 Market making on stock markets

In the previous sections, we have presented models where quotes can be posted at any price. We have claimed that these models are ill-suited for market making on stock markets, except maybe as far as small-tick stocks are concerned. In this section, we present a discrete model for market making on the stock market, which is inspired by the model of Guilbaud and Pham [92]. In this model – similar to the optimal execution model presented in Chapter 7 – market making is tackled in the general sense of trading on both sides: the trader can post limit orders on both sides of the book, together with market orders, and his goal is to make money out of the difference between the prices of his buy and sell limit orders.

Like in Chapter 7, we consider a simplified limit order book modeled by two processes: a martingale process $(S_t)_t$ for the mid-price – with infinitesimal generator $\mathcal{L}$ – and a stochastic process $(\psi_t)_t$ for the bid-ask spread, assumed to be a continuous-time Markov chain taking values in $\{\delta, \ldots, J\delta\}$. We denote by $(r_{j\delta,j'\delta}^{\text{spread}})_{j,j'}$ the instantaneous transition matrix of the process $(\psi_t)_t$.

In addition to the bid-ask spread, other indicators are introduced in order to explain the probability of execution of limit orders on each side of the limit order book, and at the different possible prices. These indicators are regarded as a single variable denoted by I. We assume that $(I_t)_t$ is a continuous-time Markov chain taking its values in a discrete set $\{I_1, \ldots, I_K\}$. We denote by $(r_{k,k'}^{\text{ind}})_{k,k'}$ the instantaneous transition matrix of the process $(I_t)_t$.[32]

We consider that limit orders all have the same size l: we write $l_t^b = l$ (respectively $l_t^a = l$) when a limit order is posted on the bid (respectively ask) side at time t, and $l_t^b = 0$ (respectively $l_t^a = 0$) otherwise. Buy limit orders can be posted at the best bid, or at one tick from the best bid, either to improve the best bid price, or at the second limit to get a high priority in the queue.

---

[32]Like in Chapter 7, we assume that the processes $(\psi_t)_t$ and $(I_t)_t$ are independent. Nonetheless, it is straightforward to generalize the model, by considering a transition matrix for the process $(\psi_t, I_t)_t$.

Similarly, sell limit orders can be posted at the best ask, or at one tick from the best ask, either to improve the best ask price, or at the second limit to get a high priority in the queue. The execution[33] of limit orders is modeled by two point processes: $(N_t^b)_t$ for the number of buy limit orders executed and $(N_t^a)_t$ for the number of sell limit orders executed. The intensities of these processes are respectively given by $\lambda^b(\psi_t, I_t, p_t^b)$ and $\lambda^a(\psi_t, I_t, p_t^a)$, where $S_t - \frac{\psi_t}{2} - p_t^b\delta$ and $S_t + \frac{\psi_t}{2} + p_t^a\delta$ are respectively the price of the buy and sell limit orders that have been posted ($p_t^b, p_t^a \in \{0, 1\}$ if $\psi_t = \delta$, and $p_t^b, p_t^a \in \{-1, 0, 1\}$ otherwise).[34]

In addition to limit orders, market orders can be sent on both sides.[35] We denote by J^b and J^a the jump processes corresponding to sell and buy market orders, i.e., $J_t^b = J_{t-}^b + 1$ when a sell market order is sent at time t, and $J_t^a = J_{t-}^a + 1$ when a buy market order is sent at time t. We assume that the sizes of market orders are respectively m_t^b and m_t^a (we choose these sizes in the interval $[0, \overline{m}]$), and we assume that the execution price is $S_t - \frac{\psi_t}{2}$ for sell orders and $S_t + \frac{\psi_t}{2}$ for buy orders.[36]

The resulting dynamics for the number q of shares in the portfolio is

$$dq_t = l_t^b dN_t^b + m_t^a dJ_t^a - l_t^a dN_t^a - m_t^b dJ_t^b.$$

The associated dynamics for the cash account is given by

$$\begin{aligned} dX_t &= l_t^a \left(S_t + \frac{\psi_t}{2} + p_t^a\delta \right) dN_t^a + m_t^b \left(S_t - \frac{\psi_t}{2} \right) dJ_t^b \\ &\quad - l_t^b \left(S_t - \frac{\psi_t}{2} - p_t^b\delta \right) dN_t^b - m_t^a \left(S_t + \frac{\psi_t}{2} \right) dJ_t^a. \end{aligned}$$

Unlike what we have done in the previous sections, we do not consider an expected utility framework with a CARA utility function. Instead, we consider the expected value of the P&L minus a term that penalizes large (long or short) inventories:

$$\mathbb{E}\left[X_T + q_T S_T - C \int_0^T q_t^2 dt \right],$$

where C may be a proxy for risk aversion and/or for the volatility of the stock price.

---

[33]There is no partial fill in the model.

[34]We authorize $p_t^b = p_t^a = -1$, even when $\psi_t = 2\delta$, but this is not a real issue because it is equivalent to sending no limit order.

[35]The model can be improved, with no additional difficulty, to take into account trading fees and rebates.

[36]$\overline{m}$ has to be small for this hypothesis to be relevant.

In order to solve this problem, we introduce the value function $u(t, x, q, S, \psi, I)$. It is viscosity solution of the following QVI:

$$
\begin{aligned}
0 = \min\Bigg\{ & -\partial_t u(t, x, q, S, \psi, I) - \mathcal{L}u(t, x, q, S, \psi, I) + Cq^2 \\
& -\sum_{j=1}^{J} r_{\psi, j\delta}^{\text{spread}} \left(u(t, x, q, S, j\delta, I) - u(t, x, q, S, \psi, I)\right) \\
& -\sum_{k=1}^{K} r_{I, I_k}^{\text{ind}} \left(u(t, x, q, S, \psi, I_k) - u(t, x, q, S, \psi, I)\right) \\
& -\left(\sup_{p^b \in Q_\psi} \lambda(\psi, I, p^b) \left(u\left(t, x - l\left(S - \frac{\psi}{2} - p^b\delta\right), q+l, S, \psi, I\right) \right.\right. \\
& \qquad\qquad\qquad \left.\left. -u(t, x, q, S, \psi, I) \right)\right)_+ \\
& -\left(\sup_{p^a \in Q_\psi} \lambda(\psi, I, p^a) \left(u\left(t, x + l\left(S + \frac{\psi}{2} + p^a\delta\right), q-l, S, \psi, I\right) \right.\right. \\
& \qquad\qquad\qquad \left.\left. -u(t, x, q, S, \psi, I) \right)\right)_+ ; \\
& u(t, x, q, S, \psi, I) - \sup_{m^a \in [0, \overline{m}]} u\left(t, x - m^a\left(S + \frac{\psi}{2}\right), q + m^a, S, \psi, I\right) ; \\
& u(t, x, q, S, \psi, I) - \sup_{m^b \in [0, \overline{m}]} u\left(t, x + m^b\left(S - \frac{\psi}{2}\right), q - m^b, S, \psi, I\right) \Bigg\}
\end{aligned}
$$

$$(11.27)$$

where $Q_\psi = \{0, 1\}$ if $\psi = \delta$, and $Q_\psi = \{-1, 0, 1\}$ otherwise.

The terminal condition is

$$u(T, x, q, S, \psi, I) = x + qS.$$

This equation is a bit complicated, but it can be simplified through a simple change of variables. In order to compute u and the optimal strategy, we use indeed the same ansatz as in Chapter 7.

This leads to

$$u(t, x, q, S, \psi, I) = x + qS + \phi_{\psi, I}(t, q).$$

It is then straightforward to verify that $(\phi_{\psi,I})_{\psi,I}$ is viscosity solution of

$$
\begin{aligned}
0 = \min \Bigg\{ & -\partial_t \phi_{\psi,I}(t,q) + Cq^2 - \sum_{j=1}^{J} r_{\psi,j\delta}^{\text{spread}} \left(\phi_{j\delta,I}(t,q) - \phi_{\psi,I}(t,q) \right) \\
& - \sum_{k=1}^{K} r_{I,I_k}^{\text{ind}} \left(\phi_{\psi,I_k}(t,q) - \phi_{\psi,I}(t,q) \right) \\
& - \left(\sup_{p^a \in Q_\psi} \lambda(\psi,I,p^b) \left(l\left(\frac{\psi}{2} + p^b \delta\right) + \phi_{\psi,I}(t,q+l) - \phi_{\psi,I}(t,q) \right) \right)_+ \\
& - \left(\sup_{p^a \in Q_\psi} \lambda(\psi,I,p^a) \left(l\left(\frac{\psi}{2} + p^a \delta\right) + \phi_{\psi,I}(t,q-l) - \phi_{\psi,I}(t,q) \right) \right)_+ ; \\
& \phi_{\psi,I}(t,q) - \sup_{m^a \in [0,\overline{m}]} \left(-m^a \frac{\psi}{2} + \phi_{\psi,I}(t,q+m^a) \right) ; \\
& \phi_{\psi,I}(t,q) - \sup_{m^b \in [0,\overline{m}]} \left(-m^b \frac{\psi}{2} + \phi_{\psi,I}(t,q-m^b) \right) \Bigg\},
\end{aligned}
\tag{11.28}
$$

with terminal condition $\phi_{\psi,I}(T,q) = 0$.

Like in the model of Chapter 7, the complex quasi-variational inequality (11.27) involves a function of 6 variables. By using $(\phi_{\psi,I})_{\psi,I}$, the problem boils down to the 2D system of equations (11.28) involving functions of 2 variables, for which a solution can be numerically approximated very easily on a (t,q,ψ,I)-grid – see [92] for a simpler case. Therefore, we can compute, and tabulate in advance, the optimal decision as a function of the time variable t, the current inventory q, and the state of the trading environment, represented by the couple (ψ, I). As in most market making models, the time variable t only matters because of the final time T. In particular, the numerical computations can be stopped whenever optimal behaviors stabilize in the backward induction process.

Like in Chapter 7, we think that some stochastic optimal control models of this kind could be used in practice, especially if they are generalized to account for the existence of several venues on which it is possible to trade the same stock.

Stochastic optimal control theory is very powerful for addressing market making issues, but machine learning techniques should also be considered to build optimal trading models. We bet that the Quantitative Finance research community will rely more and more on machine learning techniques in the near future, especially reinforcement learning techniques.

11.5 Conclusion

Market making models are often very similar to those dealing with optimal execution. In this chapter we have seen that the Avellaneda-Stoikov model – which has inspired the first optimal execution models with limit orders (see Chapter 7) – is a very interesting model for understanding the trade-off faced by market makers. This model can also be generalized in order to be applied to many quote-driven markets. The corporate bond market is one instance of a market in which models similar to that of Section 11.3 are used by practitioners for automatically streaming quotes, and even sending executable quotes.

As far as market making on stock markets is concerned, discrete models should be favored; the model presented in Section 11.4 is one simple example of a discrete model that could be used in practice – successfully if one chooses the right indicators.

Market making on derivatives markets is not addressed in this book. Market making on option markets is in fact one of the current hot topics in academic research. Models are obviously more complicated than those presented in this chapter because optimal strategies involve both the options and the underlying asset(s). We refer the interested reader to [83] and [164].

This chapter is the final one. We hope that the journey we have proposed, from optimal execution to market making issues, was interesting and thought-provoking for the reader.

Mathematical Appendices

Appendix A

Mathematical economics

If people do not believe that mathematics is simple, it is only because they do not realize how complicated life is.

— John von Neumann

In this appendix, we recall a few concepts of mathematical economics. In particular, we focus on utility functions, and their use for dealing with risk.

A.1 The expected utility theory

Should one prefer a low-risk low-return strategy or a high-risk high-return strategy? This question echoes the heated debates that have occurred between scientists over the last centuries. Some major scientists such as Cramer, N. Bernoulli (who originally introduced what is now called the St. Petersburg paradox), or his cousin D. Bernoulli tackled the general question of optimal choice under uncertainty. In particular, the latter introduced the idea of marginal utility and set the basis of the expected utility theory that was formally developed by von Neumann and Morgenstern in the middle of the 20th century.

Von Neumann and Morgenstern showed that if the preferences of an agent regarding random lotteries verify 4 natural axioms (see [147]), then their preferences can be represented by a utility function. In other words, there exists an increasing and concave function u – called a utility function – such that a lottery X (that is, a real-valued random variable modeling the outcome of the gamble) is preferred to a lottery Y if and only if $\mathbb{E}[u(X)] \geq \mathbb{E}[u(Y)]$.

The utility function representing the preferences of such a rational agent is defined up to an affine and increasing transformation. In other words, the utility functions u and $au + b$ $(a > 0)$ represent the same preferences.

A.2   Utility functions and risk aversion

Utility functions have been introduced for going beyond choices based on the expected value of lotteries. In fact, the concavity of a utility function u makes it possible to take account of the risk. This is formalized in the following proposition:

Proposition A.1. *Let u be a utility function (i.e., increasing and concave). Then, $\forall X \in L^1(\Omega)$, such that $u(X) \in L^1(\Omega)$, we have*

$$\mathbb{E}[u(X)] \leq u(\mathbb{E}[X]).$$

This proposition – which is a simple consequence of Jensen's inequality for concave functions – states that a rational agent would rather get the expected value of a lottery than play the lottery.

When $u(x) = ax + b, a > 0$, we say that the agent is risk neutral. Otherwise, the agent is risk averse. To quantify risk aversion, we use the absolute risk aversion function:

Definition A.1. *Let u be a twice differentiable utility function such that $u' > 0$. We define the absolute risk aversion function $\gamma(\cdot)$ as*

$$\gamma(x) = -\frac{u''(x)}{u'(x)}.$$

This function is nonnegative and it is invariant by increasing affine transformation of u.

A class of utility functions plays a specific role: the constant absolute risk aversion (CARA) utility functions. A twice differentiable utility function is CARA if and only if the associated absolute risk aversion function $\gamma(\cdot)$ is a constant function (the constant being denoted by γ). CARA utility functions are of the form

$$u(x) = -a \exp(-\gamma x) + b, a > 0,$$

for $\gamma > 0$, and

$$u(x) = ax + b, a > 0,$$

for $\gamma = 0$.

In practice, we often consider $u(x) = -\exp(-\gamma x)$ or $u(x) = x$, the latter function being the limit of $\frac{1 - \exp(-\gamma x)}{\gamma}$ when γ tends to 0.

Other classes of utility functions play an important role in economics and finance, such as the constant relative risk aversion (CRRA) utility functions.

These CRRA utility functions are of the form

$$u(x) = a\frac{x^{1-\rho}}{1-\rho} + b, \quad a > 0,$$

where $\rho \in [0,1) \cup (1, +\infty)$, or

$$u(x) = a\log(x) + b, \quad a > 0,$$

which corresponds to the limit $\rho \to 1$.[1]

A.3 Certainty equivalent and indifference pricing

To quantify risk aversion, another notion plays a major role in economics: the notion of certainty equivalent.

Definition A.2. *Let u be a continuous and increasing utility function. Let X be a real-valued random variable such that X and $u(X)$ are in $L^1(\Omega)$. The certainty equivalent of X for the utility function u is the unique $e \in \mathbb{R}$ such that*

$$\mathbb{E}[u(X)] = u(e).$$

The uniqueness of e comes from the fact that u is increasing.

The certainty equivalent of a lottery is the risk-free payoff that has the same expected utility as the lottery. By using the inequality of Proposition A.1 and the monotony of u, we get $e \leq \mathbb{E}(X)$.

The certainty equivalent is often approximated by using the so-called Arrow-Pratt approximation formula[2]

$$e \simeq \mathbb{E}[X] - \frac{\gamma(\mathbb{E}[X])}{2}\mathbb{V}[X].$$

---

[1]These functions are defined on $\mathbb{R}$ by setting the value $-\infty$ on the half line $x < 0$. They correspond to $\gamma(x) = \frac{\rho}{x}$. CRRA utility functions are not used in this book. We only consider utility functions defined on $\mathbb{R}$ and taking finite values.

[2]On the one hand we have

$$\mathbb{E}[u(X)] \simeq \mathbb{E}\left[u(\mathbb{E}[X]) + u'(\mathbb{E}[X])(X - \mathbb{E}[X]) + \frac{1}{2}u''(\mathbb{E}[X])(X - \mathbb{E}[X])^2\right]$$

$$\simeq u(\mathbb{E}[X]) + \frac{1}{2}u''(\mathbb{E}[X])\mathbb{V}[X].$$

On the other hand

$$u(e) \simeq u(\mathbb{E}[X]) + u'(\mathbb{E}[X])(e - \mathbb{E}[X]).$$

By definition of e we have therefore

$$u(\mathbb{E}[X]) + \frac{1}{2}u''(\mathbb{E}[X])\mathbb{V}[X] \simeq u(\mathbb{E}[X]) + u'(\mathbb{E}[X])(e - \mathbb{E}[X]),$$

or equivalently

$$e \simeq \mathbb{E}[X] - \frac{\gamma(\mathbb{E}[X])}{2}\mathbb{V}[X].$$

In finance, we often use the concept of indifference price, which is very close to the concept of certainty equivalent.

Definition A.3. *Let u be a continuous and increasing utility function. Let X be a real-valued random variable such that X and $u(X)$ are in $L^1(\Omega)$. Let $x \in \mathbb{R}$. The indifference price associated with the lottery X, when the agent has initial wealth x, is the unique $p(x) \in \mathbb{R}$, such that*

$$\mathbb{E}[u(x + X - p(x))] = u(x).$$

In other words, the indifference price associated with a lottery is the maximum price the agent is ready to pay to play the lottery. It depends on the initial wealth x, except in the case of CARA utility functions (this is stated in Proposition A.2).

Proposition A.2. *Let u be a CARA utility function. Then the indifference price of a lottery is the certainty equivalent of this lottery. In particular, it is independent of the initial wealth.*

Proof. *Let γ be the constant absolute risk aversion coefficient associated with u. Then, there exist $(a, b) \in \mathbb{R}_+ \times \mathbb{R}$, such that*

$$\forall z \in \mathbb{R}, u(z) = -a\exp(-\gamma z) + b.$$

By definition we have

$$
\begin{aligned}
\mathbb{E}[u(x + X - p(x))] = u(x) &\iff -a\mathbb{E}\left[e^{-\gamma(x + X - p(x))}\right] + b = -ae^{-\gamma x} + b \\
&\iff e^{-\gamma x}e^{\gamma p(x)}\mathbb{E}\left[e^{-\gamma X}\right] = e^{-\gamma x} \\
&\iff \mathbb{E}\left[e^{-\gamma X}\right] = e^{\gamma p(x)} \\
&\iff \mathbb{E}\left[u(X)\right] = u(p(x)).
\end{aligned}
$$

Therefore, $p(x)$ is the certainty equivalent of X. In particular, it is independent of x.

One last point, which is used recurrently in this book, is the closed-form expression for the certainty equivalent/indifference price of a Gaussian lottery, in the case of a CARA utility function. This is related to the Laplace transform of a Gaussian variable.

Theorem A.1. *Let X be a real-valued Gaussian random variable $\mathcal{N}(\mu, \sigma^2)$. Let $\gamma > 0$.*

$$\mathbb{E}[-\exp(-\gamma X)] = -\exp\left(-\gamma\mu + \frac{1}{2}\gamma^2\sigma^2\right).$$

*In other words, the certainty equivalent/indifference price associated with X
when the utility function is CARA with absolute risk aversion* γ, *is*

$$e = p = \mu - \frac{1}{2}\gamma\sigma^2.$$

Proof. *We have*

$$
\begin{aligned}
\mathbb{E}[-\exp(-\gamma X)] &= -\frac{1}{\sqrt{2\pi}}\int_{\mathbb{R}}\exp(-\gamma\mu-\gamma\sigma x)\exp\left(-\frac{x^2}{2}\right)dx \\
&= -\frac{1}{\sqrt{2\pi}}\exp(-\gamma\mu)\int_{\mathbb{R}}\exp\left(-\frac{1}{2}(x+\gamma\sigma)^2+\frac{1}{2}\gamma^2\sigma^2\right)dx \\
&= -\exp\left(-\gamma\mu+\frac{1}{2}\gamma^2\sigma^2\right)\frac{1}{\sqrt{2\pi}}\int_{\mathbb{R}}\exp\left(-\frac{1}{2}(x+\gamma\sigma)^2\right)dx \\
&= -\exp\left(-\gamma\mu+\frac{1}{2}\gamma^2\sigma^2\right).
\end{aligned}
$$

Theorem A.1 says that the Arrow-Pratt approximation is in fact exact in
the case of a Gaussian variable and a CARA utility function.

Appendix B

Convex analysis and variational calculus

Mathematics possesses not only truth, but supreme beauty.

— Bertrand Russell

In this book, we often rely on theoretical results of convex analysis, especially convex duality. In this appendix, we present the main results we need, with a particular focus on the notion of subdifferentiability, and on the Legendre-Fenchel transform of a convex function. For most proofs on convex functions, we refer the reader to undergraduate courses dealing with convex functions, or to classical books of convex analysis such as [156]. We decided to present the proofs for the results related to the Legendre-Fenchel transform, as we often use conjugate functions in this book.

Convex analysis is often used in this book to solve optimization problems. In this appendix, we recall a few results of variational calculus, especially as far as Bolza problems are concerned. In addition to recalling the classical Euler-Lagrange equation and the Hamiltonian system characterizing the solution of a Bolza problem in continuous time, we address in detail the case of discrete-time problems (that are far less classical than their continuous-time counterparts).

B.1 Basic notions of convex analysis

B.1.1 Definitions and classical properties

In what follows, we consider C a convex set of the Euclidian space $\mathbb{R}^d$ for $d \geq 1$.

We recall the definition of a convex function on C:

Definition B.1. *A function $f : C \to \mathbb{R}$ is convex if and only if*

$$\forall x, y \in C, \forall t \in [0, 1], f(tx + (1-t)y) \leq tf(x) + (1-t)f(y).$$

In particular, if

$$\forall x, y \in C, \forall t \in (0, 1), f(tx + (1 - t)y) < tf(x) + (1 - t)f(y),$$

then we say that the function is strictly convex.

A related concept is concavity. We say that $f : C \to \mathbb{R}$ is concave, if $-f$ is convex.

Except maybe at the boundary of its domain of definition, a convex function is smooth. This is the purpose of the following proposition:

Proposition B.1. *Let $f : C \to \mathbb{R}$ be a convex function. Then f is locally Lipschitz on the interior of C (denoted by $\mathring{C}$). In particular, f is continuous on $\mathring{C}$.*

Now, as far as differentiability is concerned, we have:

Proposition B.2. *Let $f : C \to \mathbb{R}$ be a convex function. Then f is almost everywhere differentiable on $\mathring{C}$. Moreover,*

$$\forall x \in \mathring{C}, \forall v \in \mathbb{R}^d, \lim_{t \to 0, t > 0} \frac{f(x + tv) - f(x)}{t} \text{ exists and is finite.}$$

Although it has directional derivatives, a convex function may not be differentiable. However, when a function is differentiable on an open set, its gradient is automatically continuous. This is stated in the following proposition.

Proposition B.3. *Let $f : C \to \mathbb{R}$ be a convex function. If f is differentiable on $\mathring{C}$, then $f \in C^1(\mathring{C})$.*

B.1.2 Subdifferentiability

When a convex function is not differentiable, a weaker notion is used, that plays a similar role as differentiability in convex optimization problems: subdifferentiability.

Definition B.2. *Let $f : C \to \mathbb{R}$ be a convex function. Let $x \in C$. $p \in \mathbb{R}^d$ is called a subgradient of f at point x if and only if*

$$\forall y \in C, f(y) \geq f(x) + p \cdot (y - x).$$

The subdifferential of f at point x, denoted by $\partial^- f(x)$, is the set of such subgradients:

$$\partial^- f(x) = \left\{ p \in \mathbb{R}^d, \forall y \in C, f(y) \geq f(x) + p \cdot (y - x) \right\}.$$

A function is said to be subdifferentiable at point x if $\partial^- f(x) \neq \emptyset$.

A convex function may not be subdifferentiable at the boundary of its domain, but its subdifferential is nonempty at any point of the interior of its domain. The link between subdifferentiability and differentiability is given in the following proposition:

Proposition B.4. *Let $f : C \to \mathbb{R}$ be a convex function. Let $x \in \mathring{C}$. The following two assertions are equivalent:*

- *f is differentiable at x,*
- *$\partial^- f(x)$ is a singleton.*

If $\partial^- f(x)$ is a singleton, then necessarily $\partial^- f(x) = \{\nabla f(x)\}$.

The subdifferential of a convex function f is often used to find the minimizers of a convex function. The following proposition generalizes the classical first-order condition to find the minimizers of a differentiable convex function:

Proposition B.5. *Let $f : C \to \mathbb{R}$ be a convex function. $x \in C$ is a minimizer of f if and only if $0 \in \partial^- f(x)$.*

B.1.3 The Legendre-Fenchel transform

We now introduce a central tool of convex analysis, which is used recurrently in this book: the Legendre-Fenchel transform of a convex function.

In this section, we cover the case of classical convex functions with super-linear growth, for which proofs are easily obtained. Then we shall explain how the results generalize to the case of generalized convex functions that can take the value $+\infty$.

In what follows, we consider a convex function $f : \mathbb{R}^d \to \mathbb{R}$, assumed to be asymptotically super-linear, i.e., $\lim_{|x|\to+\infty} \frac{f(x)}{|x|} = +\infty$.

Definition B.3. *The Legendre-Fenchel transform of f is defined by*

$$f^* : p \in \mathbb{R}^d \mapsto \sup_{x \in \mathbb{R}^d} p \cdot x - f(x).$$

This function is also called the conjugate function of f, or simply the conjugate of f.

f^* is well defined (it only takes finite values) thanks to the assumption of super-linearity.

The convex conjugate of a function f is usually denoted by f^*. However, when the initial convex function is denoted by L (for Lagrangian), L^* is often denoted by H (for Hamiltonian).

Proposition B.6. *f^* verifies the following properties:*

- *f^* is a convex function,*
- *f^* is asymptotically super-linear.*

Proof. *Let $(p_1, p_2) \in \mathbb{R}^d \times \mathbb{R}^d, t \in [0,1], x \in \mathbb{R}^d,$*

$$
\begin{aligned}
(tp_1 + (1-t)p_2) \cdot x - f(x) &= t(p_1 \cdot x - f(x)) + (1-t)(p_2 \cdot x - f(x)) \\
&\leq tf^*(p_1) + (1-t)f^*(p_2).
\end{aligned}
$$

Therefore,

$$
f^*(tp_1 + (1-t)p_2) \leq tf^*(p_1) + (1-t)f^*(p_2).
$$

To prove the second point, we simply write, for $p \in \mathbb{R}^d \setminus \{(0, \ldots, 0)\}$, and $M > 0$:

$$
f^*(p) \geq p \cdot M\frac{p}{|p|} - f\left(M\frac{p}{|p|}\right) \geq M|p| - \max_{|x|=M} f(x).
$$

Therefore,

$$
\liminf_{|p| \to +\infty} \frac{f^*(p)}{|p|} \geq M.
$$

This being true for all $M > 0$, we have

$$
\lim_{|p| \to +\infty} \frac{f^*(p)}{|p|} = +\infty.
$$

Proposition B.6 allows us to define the biconjugate $f^{**} = (f^*)^*$ of f. The notion of biconjugate is at the heart of a central (and one of the most powerful) theorem of convex analysis.

Theorem B.1. *The biconjugate f^{**} of the function f is the function f itself. Furthermore,*

$$
\forall (x, p) \in \mathbb{R}^d \times \mathbb{R}^d, p \in \partial^- f(x) \iff x \in \partial^- f^*(p).
$$

Proof. *By definition, we have*

$$
\forall (x, p) \in \mathbb{R}^d \times \mathbb{R}^d, f^*(p) \geq p \cdot x - f(x).
$$

This also writes

$$
\forall (x, p) \in \mathbb{R}^d \times \mathbb{R}^d, f(x) \geq p \cdot x - f^*(p).
$$

Therefore, $f \geq f^{**}$.

To prove the converse inequality, let us consider $x \in \mathbb{R}^d$. We consider p a subgradient of f at x, i.e., $p \in \partial^- f(x)$.

By definition of p, $\forall y \in \mathbb{R}^d$, $f(y) \geq f(x) + p \cdot (y - x)$. Therefore, x maximizes the function $y \mapsto p \cdot y - f(y)$. This also writes

$$f^*(p) = p \cdot x - f(x).$$

Therefore,

$$f(x) = p \cdot x - f^*(p) \leq f^{**}(x).$$

We conclude that $f \leq f^{**}$, and therefore that $f = f^{**}$.

Now, $\forall (x, p) \in \mathbb{R}^d \times \mathbb{R}^d$,

$$p \in \partial^- f(x)$$
$$\Longleftrightarrow \quad f^*(p) = p \cdot x - f(x)$$
$$\Longleftrightarrow \quad f(x) = p \cdot x - f^*(p)$$
$$\Longleftrightarrow \quad f^{**}(x) = p \cdot x - f^*(p)$$
$$\Longleftrightarrow \quad x \in \partial^- f^*(p).$$

One last result, which is often used in this book, is the following:

Proposition B.7. *If f is strictly convex, then f^* is continuously differentiable.*

Proof. *Because of Proposition B.3, we only need to prove that f^* is differentiable on $\mathbb{R}^d$.*

We proceed by contradiction. If $p \in \mathbb{R}^d$ is such that f^* is not differentiable at p, then there exist two distinct points x_1 and x_2 in $\partial^- f^*(p)$. Using Theorem B.1, we have

$$p \in \partial^- f(x_1) \cap \partial^- f(x_2).$$

By definition, this gives $\forall x \in \mathbb{R}^d$,

$$f(x) \geq f(x_1) + p(x - x_1), \tag{B.1}$$
$$f(x) \geq f(x_2) + p(x - x_2). \tag{B.2}$$

In particular, by setting $x = x_2$ in Eq. (B.1) and $x = x_1$ in Eq. (B.2), we get

$$f(x_2) \geq f(x_1) + p(x_2 - x_1), \qquad f(x_1) \geq f(x_2) + p(x_1 - x_2).$$

Therefore,

$$f(x_2) = f(x_1) + p(x_2 - x_1).$$

Now, if $t \in [0, 1]$, we set $x = t x_1 + (1 - t) x_2$ in Eq. (B.1), and we get

$$f(t x_1 + (1 - t) x_2) \geq f(x_2) + t p(x_1 - x_2)$$
$$\geq f(x_2) + t(f(x_1) - f(x_2))$$
$$\geq t f(x_1) + (1 - t) f(x_2).$$

Since f is convex, this means that

$$f(tx_1 + (1-t)x_2) = tf(x_1) + (1-t)f(x_2), \forall t \in [0,1].$$

This contradicts the strict convexity of f.

B.1.4   Generalized convex functions

In the previous paragraphs, we have only considered the case of convex functions taking finite values. Sometimes in this book, we also use a general form of convex functions taking the value $+\infty$. In particular, allowing convex functions to take the value $+\infty$ makes it possible to consider optimization problems under constraints (the constraints being incorporated in the definition of the objective function to minimize).

In this book, a function $f : \mathbb{R}^d \to \mathbb{R} \cup \{+\infty\}$ is called a generalized convex function[1] if the following conditions hold:

- f is not identically equal to $+\infty$,

- $\forall x, y \in \mathbb{R}^d, \forall t \in [0,1], f(tx + (1-t)y) \le tf(x) + (1-t)f(y)$, where the order $\le$ is naturally extended from $\mathbb{R}$ to $\mathbb{R} \cup \{+\infty\}$,

- f is lower semi-continuous in the sense that its epigraph

$$\text{epi}(f) = \{(x, \alpha) \in \mathbb{R}^d \times \mathbb{R}, \alpha \ge f(x)\}$$

  is closed.

It is noteworthy that we only consider functions defined on $\mathbb{R}^d$, not on a convex subset of $\mathbb{R}^d$. In fact, if f is defined on a convex subset C of $\mathbb{R}^d$, we extend the function f to $\mathbb{R}^d$ by setting the value $+\infty$ outside of C.

Most of the above results on convex functions generalize to generalized convex functions. As far as regularity is concerned, Propositions B.1, B.2, and B.3 apply, as soon as we consider the restriction of f to its domain $\text{dom}(f)$, defined as

$$\text{dom}(f) = \{x \in \mathbb{R}^d, f(x) < +\infty\}.$$

As far as subdifferentiability is concerned, we set $\partial^- f(x) = \emptyset, \forall x \notin \text{dom}(f)$. It is noteworthy that the subdifferential of f at any point of the boundary of $\text{dom}(f)$ may also be empty.

---

[1] This definition is not a standard definition of convex analysis. However, it covers all the cases encountered in this book.

When it comes to the Legendre-Fenchel transform of a generalized convex function f, we define f^* by

$$f^* : p \in \mathbb{R}^d \mapsto \sup_{x \in \mathbb{R}^d} p \cdot x - f(x) \in \mathbb{R} \cup \{+\infty\}.$$

f^* is a generalized convex function, and the biconjugate $f^{**} = (f^*)^*$ is therefore another generalized convex function.

Theorem B.1 holds for generalized convex functions, thanks to the lower semi-continuity assumption on f (see [156]). In particular, $f = f^{**}$ for any convex function on $\mathbb{R}^d$, independently of the asymptotic super-linearity.

When the generalized convex function f verifies the assumption of asymptotic super-linearity, it is important to notice that f^* is a convex function taking finite values only. Furthermore, the following counterpart of Proposition B.7 holds:[2]

Proposition B.8. *If f is strictly convex on dom(f), then the convex function f^* is continuously differentiable.*

B.2 Calculus of variations

B.2.1 Bolza problems in continuous time

The optimal execution problems tackled in this book often take the form of a particular class of optimization problems, called convex problems of Bolza. Here, we present a particular case of Bolza problems that is relevant for the problems encountered in this book. For a more general treatment, we refer the reader to [155, 157].

Basically, the problem is to find the minimizers of a functional of the form

$$J : x \in W^{1,1}((0,T), \mathbb{R}^d) \mapsto \int_0^T (f_t(x(t)) + g_t(x'(t))) \, dt,$$

over the set $\mathcal{C} = \{x \in W^{1,1}((0,T), \mathbb{R}^d), x(0) = a, x(T) = b\}$, where a and b are fixed constants in $\mathbb{R}^d$.

Under mild assumptions on the functions $f : (t,x) \in [0,T] \times \mathbb{R}^d \mapsto f_t(x)$ and $g : (t,v) \in [0,T] \times \mathbb{R}^d \mapsto g_t(v)$, we present differential characterizations of

---

[2]The proof is the same, because f^* is not a generalized convex function, but a classical convex function.

the minimizers, in the form of a Euler-Lagrange equation and a Hamiltonian system of equations. The framework we use in this appendix is general enough to cover all the cases found in this book.

Let us assume that f and g verify the following hypotheses (we use the same hypotheses as in [155, 157]):

- (Ha), $\forall t \in [0, T]$, $f_t(\cdot)$ and $g_t(\cdot)$ are generalized convex functions,

- (Hb), f and g are measurable with respect to the σ-field in $[0, T] \times \mathbb{R}^d$ generated by Lebesgue sets in $[0, T]$ and Borel sets in $\mathbb{R}^d$.

- (Hc), $\exists p \in L^\infty((0, T), \mathbb{R}^d), \exists s \in L^1((0, T), \mathbb{R}^d), \exists \alpha \in L^1(0, T)$, such that $f_t(x) + g_t(v) \geq x \cdot s(t) + v \cdot p(t) - \alpha(t)$, a.e.

- (Hd), $\exists x \in L^\infty((0, T), \mathbb{R}^d), \exists v \in L^1((0, T), \mathbb{R}^d), \exists \beta \in L^1(0, T)$, such that $f_t(x(t)) + g_t(v(t)) \leq \beta(t)$, a.e.

(Ha) simply states that the problem is convex, in both x and v (i.e., x'). (Hb) guarantees that the integral in the definition of J makes sense. (Hb) is true for instance if the functions f and g are lower semi-continuous. (Hc) and (Hd) are very weak assumptions in practice. For instance, (Hc) is verified if f and g are bounded from below. (Hb), (Hc), and (Hd) are also automatically verified if f and g are independent of t.

To state the differential characterizations of the minimizers of J, we introduce a dual optimization problem (the initial problem being called the primal one) which consists in minimizing

$$I : p \in W^{1,1}((0, T), \mathbb{R}^d) \mapsto \int_0^T (f_t^*(p'(t)) + g_t^*(p(t))) \, dt + a \cdot p(0) - b \cdot p(T),$$

over the set of absolutely continuous functions p.

Then, we have:

Theorem B.2. *If $x^* \in \mathcal{C}$ minimizes J over $\mathcal{C}$, and $p^* \in W^{1,1}((0, T), \mathbb{R}^d)$ minimizes I over $W^{1,1}((0, T), \mathbb{R}^d)$, then*

- *The Euler-Lagrange equation (B.3) is satisfied:*

$$\begin{cases} p^{*\prime}(t) & \in \quad \partial^- f_t(x^*(t)), \\ p^*(t) & \in \quad \partial^- g_t \left(x^{*\prime}(t) \right), \\ x^*(0) & = \quad a, \\ x^*(T) & = \quad b. \end{cases} \tag{B.3}$$

- *The Hamiltonian system of equations (B.4) is satisfied:*

$$\begin{cases} p^{*\prime}(t) & \in \quad \partial^- f_t(x^*(t)), \\ x^{*\prime}(t) & \in \quad \partial^- g_t^* \left(p^*(t) \right), \\ x^*(0) & = \quad a, \\ x^*(T) & = \quad b. \end{cases} \tag{B.4}$$

Conversely, if $x^* \in C$ and $p^* \in W^{1,1}((0,T), \mathbb{R}^d)$ constitute a solution of (B.3), or a solution of (B.4), then x^* minimizes J over C, and p^* minimizes I over $W^{1,1}((0,T), \mathbb{R}^d)$.

It is noteworthy that, if f_t is differentiable, and if g_t is strictly convex, then Eq. (B.4) boils down to a system of ordinary differential equations (thanks to Proposition B.8):

$$\begin{cases} p^{*\prime}(t) &= f_t'(x^*(t)), \\ x^{*\prime}(t) &= g_t^{*\prime}(p^*(t)), \\ x^*(0) &= a, \\ x^*(T) &= b. \end{cases} \tag{B.5}$$

The above differential characterization is very useful in practice to study the properties of the minimizers, and to approximate numerically the minimizers of J (or I).

However, Theorem B.2 does not guarantee the existence of a minimizer. When f_t and g_t are classical convex functions, we can use Theorem B.3 below to prove the existence of a minimizer to both the primal and the dual problems. However, when it comes to generalized convex functions, general results do exist, but they can hardly be stated without complex considerations of convex analysis that are out of the scope of this appendix. For the cases covered in this book, we refer the interested reader to Theorem 1 in [156].

Theorem B.3. *Assume that*

- *f and g are continuous and bounded from below,*
- *f_t and g_t are convex functions,*
- *$\lim_{|x| \to +\infty} \inf_{t \in [0,T]} \frac{f_t(x)}{|x|} = +\infty$ and $\lim_{|v| \to +\infty} \inf_{t \in [0,T]} \frac{g_t(v)}{|v|} = +\infty$.*

Then, there exist a minimizer x^* of J over C and a minimizer p^* of I over $W^{1,1}((0,T), \mathbb{R}^d)$.

B.2.2 What about discrete-time problems?

The Hamiltonian system (B.4) is part of the mathematical culture of most people with a background in mathematical optimization or Hamiltonian mechanics. However, its discrete counterpart is less famous.

Let us consider the function $J : \mathbb{R}^{d(N+1)} \to \mathbb{R}$ defined by

$$J(x_0, \ldots, x_N) = \sum_{n=1}^{N-1} f_n(x_n) + \sum_{n=0}^{N-1} g_n(x_{n+1} - x_n),$$

where $\forall n \in \{1, \ldots, N-1\}$, f_n is a convex function, and $\forall n \in \{0, \ldots, N-1\}$, g_n is a convex function.

Our goal is to find x minimizing J over

$$\mathcal{C} = \left\{ (x_0, \ldots, x_N) \in \mathbb{R}^{d(N+1)}, x_0 = a, x_N = b \right\},$$

a and b being fixed constant in $\mathbb{R}^d$.[3]

For that purpose, we introduce a function $I : \mathbb{R}^{dN} \to \mathbb{R}$ defined by

$$I(p_0, \ldots, p_{N-1}) = \sum_{n=1}^{N-1} f_n^* (p_n - p_{n-1}) + \sum_{n=0}^{N-1} g_n^* (p_n) + a \cdot p_0 - b \cdot p_{N-1}.$$

The discrete counterpart of Theorem B.2 is the following:[4]

Theorem B.4. *If $x^* = (x_0^*, \ldots, x_N^*) \in \mathcal{C}$ minimizes J over $\mathcal{C}$, and $p^* = (p_0^*, \ldots, p_{N-1}^*) \in \mathbb{R}^{dN}$ minimizes I over $\mathbb{R}^{dN}$, then we have:*

$$\begin{cases} p_n^* - p_{n-1}^* \in \partial^- f_n (x_n^*) & \forall n \in \{1, \ldots, N-1\}, \\ x_{n+1}^* - x_n^* \in \partial^- g_n^* (p_n^*) & \forall n \in \{0, \ldots, N-1\}. \end{cases} \qquad (B.6)$$

Conversely, if $x^* = (x_0^*, \ldots, x_N^*) \in \mathcal{C}$ and $p^* = (p_0^*, \ldots, p_{N-1}^*) \in \mathbb{R}^{dN}$ constitute a solution to the system (B.6), then x^* minimizes J over $\mathcal{C}$, and p^* minimizes I over $\mathbb{R}^{dN}$.

Proof. *We first introduce*

$$\begin{aligned} \varphi : \quad & \mathbb{R}^{dN} \times \mathcal{C} \longrightarrow \mathbb{R} \\ & (y, x) \mapsto \sum_{n=0}^{N-1} g_n (x_{n+1} - x_n + y_n) + \sum_{n=1}^{N-1} f_n (x_n). \end{aligned}$$

Then, we define a convex function α by

$$\alpha : y \in \mathbb{R}^N \mapsto \alpha(y) = \inf_{x \in \mathcal{C}} \varphi (y, x).$$

Now, we define β as the Legendre-Fenchel transform of α:

$$\beta (p) = \sup_{y \in \mathbb{R}^N} \left\{ \sum_{n=0}^{N-1} y_n \cdot p_n - \alpha (y) \right\}.$$

An important result is that $\beta (p) = I(p)$.

---

[3] There is no issue with existence in this discrete-time case.

[4] We consider in this section the case of convex functions. The result generalizes to generalized convex functions, but the proof is more technical.

This result is based on the following computations:

$$
\begin{aligned}
\beta(p) &= \sup_{y \in \mathbb{R}^N} \left\{ \sum_{n=0}^{N-1} y_n \cdot p_n - \inf_{x \in \mathcal{C}} \varphi(y, x) \right\} \\
&= \sup_{x \in \mathcal{C}} \sup_{y \in \mathbb{R}^N} \left\{ \sum_{n=0}^{N-1} y_n \cdot p_n - \sum_{n=0}^{N-1} g_n(x_{n+1} - x_n + y_n) - \sum_{n=1}^{N-1} f_n(x_n) \right\} \\
&= \sup_{x \in \mathcal{C}} \sup_{y \in \mathbb{R}^N} \left\{ \sum_{n=0}^{N-1} (x_{n+1} - x_n + y_n) \cdot p_n - \sum_{n=0}^{N-1} (x_{n+1} - x_n) \cdot p_n \right. \\
&\qquad\qquad \left. - \sum_{n=0}^{N-1} g_n(x_{n+1} - x_n + y_n) - \sum_{n=1}^{N-1} f_n(x_n) \right\} \\
&= \sup_{x \in \mathcal{C}} \sup_{y \in \mathbb{R}^N} \left\{ \sum_{n=0}^{N-1} y_n \cdot p_n - \sum_{n=0}^{N-1} (x_{n+1} - x_n) \cdot p_n \right. \\
&\qquad\qquad \left. - \sum_{n=0}^{N-1} g_n(y_n) - \sum_{n=1}^{N-1} f_n(x_n) \right\} \\
&= \sup_{x \in \mathcal{C}} \left\{ \sup_{y \in \mathbb{R}^N} \left\{ \sum_{n=0}^{N-1} y_n \cdot p_n - \sum_{n=0}^{N-1} g_n(y_n) \right\} \right. \\
&\qquad\qquad \left. - \sum_{n=0}^{N-1} (x_{n+1} - x_n) \cdot p_n - \sum_{n=1}^{N-1} f_n(x_n) \right\} \\
&= \sup_{x \in \mathcal{C}} \left\{ \sum_{n=0}^{N-1} g_n^*(p_n) - \sum_{n=0}^{N-1} (x_{n+1} - x_n) \cdot p_n - \sum_{n=1}^{N-1} f_n(x_n) \right\} \\
&= \sum_{n=0}^{N-1} g_n^*(p_n) + \sup_{x \in \mathcal{C}} \left\{ -\sum_{n=0}^{N-1} x_{n+1} \cdot p_n + \sum_{n=0}^{N-1} x_n \cdot p_n - \sum_{n=1}^{N-1} f_n(x_n) \right\} \\
&= \sum_{n=0}^{N-1} g_n^*(p_n) + \sup_{x \in \mathcal{C}} \left\{ -\sum_{n=1}^{N} x_n \cdot p_{n-1} + \sum_{n=0}^{N-1} x_n \cdot p_n - \sum_{n=1}^{N-1} f_n(x_n) \right\} \\
&= \sum_{n=0}^{N-1} g_n^*(p_n) + \sup_{x \in \mathcal{C}} \left\{ \sum_{n=1}^{N-1} x_n \cdot (p_n - p_{n-1}) \right. \\
&\qquad\qquad \left. + a \cdot p_0 - b \cdot p_{N-1} - \sum_{n=1}^{N-1} f_n(x_n) \right\} \\
&= \sum_{n=0}^{N-1} g_n^*(p_n) + \sum_{n=1}^{N-1} f_n^*(p_n - p_{n-1}) + a \cdot p_0 - b \cdot p_{N-1} \\
&= I(p).
\end{aligned}
$$

Let us suppose now that p^* minimizes I and x^* minimizes J over C. This means that $0 \in \partial^- \beta(p^*)$. By convex duality, this is equivalent to $p^* \in \partial^- \alpha(0)$. Therefore,

$$\forall y \in \mathbb{R}^{dN}, \alpha(y) - \alpha(0) \geq \sum_{n=0}^{N-1} p_n^* \cdot y_n.$$

By definition of α, this gives

$$\forall y \in \mathbb{R}^{dN}, \inf_{x \in C} \varphi(y,x) - \inf_{x \in C} \varphi(0,x) = \inf_{x \in C} \varphi(y,x) - \varphi(0,x^*) \geq \sum_{n=0}^{N-1} p_n^* \cdot y_n.$$

Therefore,

$$\forall (y,x) \in \mathbb{R}^{dN} \times C, \varphi(y,x) - \varphi(0,x^*) \geq \sum_{n=0}^{N-1} p_n^* \cdot y_n,$$

or equivalently

$$\sum_{n=0}^{N-1} g_n(x_{n+1} - x_n + y_n) - g_n(x_{n+1}^* - x_n^*)$$

$$+ \sum_{n=1}^{N} f_n(x_n) - f_n(x_n^*) \geq \sum_{n=0}^{N-1} p_n^* \cdot y_n. \qquad \text{(B.7)}$$

Given $k \in \{1, \ldots, N-1\}$ and $z \in \mathbb{R}^d$, we apply Eq. (B.7) to

$$x = (x_0^*, \ldots, x_{k-1}^*, z, x_{k+1}^*, \ldots, x_N^*),$$

and

$$y = (\underbrace{0, \ldots, 0}_{k-1}, x_k^* - z, z - x_k^*, \underbrace{0, \ldots, 0}_{N-k-1}).$$

Because $\forall n \in \{0, \ldots, N-1\}, x_{n+1} - x_n + y_n = x_{n+1}^* - x_n^*$, Eq. (B.7) gives

$$f_k(z) - f_k(x_k^*) \geq (p_k^* - p_{k-1}^*) \cdot (z - x_k^*).$$

Therefore,

$$\forall k \in \{1, \ldots, N-1\}, p_k^* - p_{k-1}^* \in \partial^- f_k(x_k^*). \qquad \text{(B.8)}$$

Now, given $k \in \{0, \ldots, N-1\}$ and $y \in \mathbb{R}^d$, we apply Eq. (B.7) to

$$x = (x_0^*, \ldots, x_N^*),$$

and

$$y = (\underbrace{0, \ldots, 0}_{k}, z, \underbrace{0, \ldots, 0}_{N-k-1}).$$

This gives

$$g_k \left(x_{k+1}^* - x_k^* + y \right) - g_k \left(x_{k+1}^* - x_k^* \right) \geq p_k^* \cdot y,$$

i.e., $p_k^* \in \partial^- g_k \left(x_{k+1}^* - x_k^* \right).$

Using convex duality, we conclude that

$$\forall k \in \{0, \ldots, N-1\}, x_{k+1}^* - x_k^* \in \partial^- g_k^* \left(p_k^* \right). \tag{B.9}$$

Eqs. (B.8) and (B.9) coincide with the system (B.6). Therefore, the first result of the theorem is proved.

Using convex duality, we have $\forall n \in \{0, \ldots, N-1\},$

$$g_n \left(x_{n+1} - x_n \right) + g_n^* \left(p_n \right) \geq p_n \cdot \left(x_{n+1} - x_n \right), \tag{B.10}$$

and $\forall n \in \{1, \ldots, N-1\},$

$$f_n \left(x_n \right) + f_n^* \left(p_n - p_{n-1} \right) \geq x_n \cdot \left(p_n - p_{n-1} \right). \tag{B.11}$$

Therefore,

$$
\begin{aligned}
I(p) + J(x) &\geq \sum_{n=0}^{N-1} p_n \cdot \left(x_{n+1} - x_n \right) + \sum_{n=1}^{N-1} x_n \cdot \left(p_n - p_{n-1} \right) + a \cdot p_0 - b \cdot p_{N-1} \\
&\geq \sum_{n=0}^{N-1} p_n \cdot x_{n+1} - \sum_{n=0}^{N-1} p_n \cdot x_n + \sum_{n=1}^{N-1} x_n \cdot p_n \\
&\quad - \sum_{n=1}^{N-1} x_n \cdot p_{n-1} + a \cdot p_0 - b \cdot p_{N-1} \\
&\geq 0.
\end{aligned}
$$

Now, if a couple $(x^*, p^*) \in \mathcal{C} \times \mathbb{R}^{dN}$ *is solution of the system (B.6), then the inequalities (B.10) and (B.11) are in fact equalities for this couple, and we get therefore, by using the same computations as above,*

$$I(p^*) + J(x^*) = 0.$$

Therefore,

$$\forall (x, p) \in \mathcal{C} \times \mathbb{R}^{dN}, I(p) + J(x) \geq I(p^*) + J(x^*).$$

In particular, by taking $p = p^*,$ *we see that* x^* *minimizes* J *over* $\mathcal{C}$. *Conversely, by taking* $x = x^*,$ *we see that* p^* *minimizes* I.

Bibliography

[1] F. Abergel, J.-P. Bouchaud, T. Foucault, C.-A. Lehalle, and M. Rosenbaum. *Market Microstructure: Confronting Many Viewpoints*. John Wiley & Sons, 2012.

[2] F. Abergel and R. Tachet des Combes. A nonlinear partial integro-differential equation from mathematical finance. *Discrete and Continuous Dynamical Systems, Series A*, 27(3):907–917, 2010.

[3] A. Alfonsi, A. Fruth, and A. Schied. Optimal execution strategies in limit order books with general shape functions. *Quantitative Finance*, 10(2):143–157, 2010.

[4] A. Alfonsi and A. Schied. Optimal trade execution and absence of price manipulations in limit order book models. *SIAM Journal on Financial Mathematics*, 1(1):490–522, 2010.

[5] A. Alfonsi, A. Schied, and A. Slynko. Order book resilience, price manipulation, and the positive portfolio problem. *SIAM Journal on Financial Mathematics*, 3(1):511 533, 2012.

[6] R. Almgren. Optimal execution with nonlinear impact functions and trading-enhanced risk. *Applied Mathematical Finance*, 10(1):1–18, 2003.

[7] R. Almgren. Optimal trading with stochastic liquidity and volatility. *SIAM Journal on Financial Mathematics*, 3(1):163–181, 2012.

[8] R. Almgren and N. Chriss. Value under liquidation. *Risk*, 12(12):61–63, 1999.

[9] R. Almgren and N. Chriss. Optimal execution of portfolio transactions. *Journal of Risk*, 3:5–40, 2001.

[10] R. Almgren, C. Thum, E. Hauptmann, and H. Li. Direct estimation of equity market impact. *Risk*, 18(7):58–62, 2005.

[11] M. Avellaneda, A. Levy, and A. Parás. Pricing and hedging derivative securities in markets with uncertain volatilities. *Applied Mathematical Finance*, 2(2):73–88, 1995.

265

[12] M. Avellaneda, J. Reed, and S. Stoikov. Forecasting prices from level-I quotes in the presence of hidden liquidity. *Algorithmic Finance*, 1(1):35–43, 2011.

[13] M. Avellaneda and S. Stoikov. High-frequency trading in a limit order book. *Quantitative Finance*, 8(3):217–224, 2008.

[14] L. Bachelier. *Théorie de la spéculation*. Gauthier-Villars, 1900.

[15] E. Bacry, A. Iuga, M. Lasnier, and C.-A. Lehalle. Market impacts and the life cycle of investors orders. *Available at SSRN 2532152*, 2014.

[16] M. Bardi and I. Capuzzo-Dolcetta. *Optimal control and viscosity solutions of Hamilton-Jacobi-Bellman equations*. Springer Science & Business Media, 2008.

[17] L. Bargeron, M. Kulchania, and S. Thomas. Accelerated share repurchases. *Journal of Financial Economics*, 101(1):69–89, 2011.

[18] G. Barles and M. Soner. Option pricing with transaction costs and a nonlinear Black-Scholes equation. *Finance and Stochastics*, 2(4):369–397, 1998.

[19] E. Bayraktar and M. Ludkovski. Liquidation in limit order books with controlled intensity. *Mathematical Finance*, 24(4):627–650, 2014.

[20] N. Bershova and D. Rakhlin. The non-linear market impact of large trades: Evidence from buy-side order flow. *Quantitative Finance*, 13(11):1759–1778, 2013.

[21] D. Bertsimas and A. Lo. Optimal control of execution costs. *Journal of Financial Markets*, 1(1):1–50, 1998.

[22] J. Białkowski, S. Darolles, and G. Le Fol. Improving VWAP strategies: A dynamic volume approach. *Journal of Banking & Finance*, 32(9):1709–1722, 2008.

[23] F. Black. How we came up with the option formula. *Journal of Portfolio Management*, 15(2):4–8, 1989.

[24] F. Black and P. Karasinski. Bond and option pricing when short rates are lognormal. *Financial Analysts Journal*, 47(4):52–59, 1991.

[25] F. Black and R. Litterman. Asset allocation: Combining investor views with market equilibrium. *The Journal of Fixed Income*, 1(2):7–18, 1991.

[26] F. Black and M. Scholes. The pricing of options and corporate liabilities. *The Journal of Political Economy*, 81(3):637–654, 1973.

[27] A. Bladon, E. Moro, and T. Galla. Agent-specific impact of single trades in financial markets. *Physical Review E*, 85(3):036103, 2012.

[28] B. Bouchard and N.-M. Dang. Generalized stochastic target problems for pricing and partial hedging under loss constraints – application in optimal book liquidation. *Finance and Stochastics*, 17(1):31–72, 2013.

[29] B. Bouchard, N.-M. Dang, and C.-A. Lehalle. Optimal control of trading algorithms: A general impulse control approach. *SIAM Journal on Financial Mathematics*, 2(1):404–438, 2011.

[30] B. Bouchard, G. Loeper, and Y. Zou. Almost-sure hedging with permanent price impact. *arXiv preprint arXiv:1503.05475*, 2015.

[31] J.-P. Bouchaud, Y. Gefen, M. Potters, and M. Wyart. Fluctuations and response in financial markets: The subtle nature of random price changes. *Quantitative Finance*, 4(2):176–190, 2004.

[32] J.-P. Bouchaud, J. Kockelkoren, and M. Potters. Random walks, liquidity molasses and critical response in financial markets. *Quantitative Finance*, 6(2):115–123, 2006.

[33] A. Brace, D. Gatarek, and M. Musiela. The market model of interest rate dynamics. *Mathematical Finance*, 7(2):127–155, 1997.

[34] X. Brokmann, J. Kockelkoren, J.-P. Bouchaud, and E. Serie. Slow decay of impact in equity markets. *Available at SSRN 2471528*, 2014.

[35] P. Cannarsa and C. Sinestrari. *Semiconcave functions, Hamilton-Jacobi equations, and optimal control*, volume 58. Springer Science & Business Media, 2004.

[36] A. Carollo, G. Vaglica, F. Lillo, and R. Mantegna. Trading activity and price impact in parallel markets: SETS vs. off-book market at the London Stock Exchange. *Quantitative Finance*, 12(4):517–530, 2012.

[37] Á. Cartea, R. Donnelly, and S. Jaimungal. Algorithmic trading with model uncertainty. *Available at SSRN 2310645*, 2013.

[38] Á. Cartea and S. Jaimungal. Risk metrics and fine tuning of high frequency trading strategies. *Mathematical Finance*, 25(3):576–611, 2013.

[39] Á. Cartea and S. Jaimungal. A closed-form execution strategy to target VWAP. *Available at SSRN 2542314*, 2014.

[40] Á. Cartea and S. Jaimungal. Incorporating order-flow into optimal execution. *Available at SSRN 2557457*, 2015.

[41] Á. Cartea and S. Jaimungal. Optimal execution with limit and market orders. *Quantitative Finance*, 15(8):1–13, 2015.

[42] Á. Cartea and S. Jaimungal. Order-flow and liquidity provision. *Available at SSRN 2553154*, 2015.

[43] Á. Cartea, S. Jaimungal, and D. Kinzebulatov. Algorithmic trading with learning. *Available at SSRN 2373196*, 2014.

[44] Á. Cartea, S. Jaimungal, and J. Penalva. *Algorithmic and High-Frequency Trading.* Cambridge University Press, 2015.

[45] Á. Cartea, S. Jaimungal, and J. Ricci. Buy low, sell high: A high frequency trading perspective. *SIAM Journal on Financial Mathematics*, 5(1):415–444, 2014.

[46] U. Çetin, R. Jarrow, and P. Protter. Liquidity risk and arbitrage pricing theory. *Finance and Stochastics*, 8(3):311–341, 2004.

[47] U. Çetin, M. Soner, and N. Touzi. Option hedging for small investors under liquidity costs. *Finance and Stochastics*, 14(3):317–341, 2010.

[48] L. Chan and J. Lakonishok. The behavior of stock prices around institutional trades. *The Journal of Finance*, 50(4):1147–1174, 1995.

[49] T. Chemmanur, Y. Cheng, and T. Zhang. Why do firms undertake accelerated share repurchase programs? *Available at SSRN 1570842*, 2010.

[50] R. Cont and A. de Larrard. Order book dynamics in liquid markets: Limit theorems and diffusion approximations. *Available at SSRN 1757861*, 2011.

[51] R. Cont and A. Kukanov. Optimal order placement in limit order markets. *arXiv preprint arXiv:1210.1625*, 2012.

[52] R. Cont, A. Kukanov, and S. Stoikov. The price impact of order book events. *Journal of Financial Econometrics*, 12(1):47–88, 2014.

[53] R. Cont and L. Wagalath. Impact of large institutional investors on the dependence structure of asset returns. Working paper, 2013.

[54] J. Cox, J. Ingersoll, and S. Ross. A theory of the term structure of interest rates. *Econometrica*, 53(2):385–407, 1985.

[55] J. Cox, S. Ross, and M. Rubinstein. Option pricing: A simplified approach. *Journal of Financial Economics*, 7(3):229–263, 1979.

[56] G. Curato, J. Gatheral, and F. Lillo. Optimal execution with nonlinear transient market impact. *Available at SSRN 2539240*, 2014.

[57] J. Cvitanić, H. Pham, and N. Touzi. A closed-form solution to the problem of super-replication under transaction costs. *Finance and Stochastics*, 3(1):35–54, 1999.

[58] K. Dayri and M. Rosenbaum. Large tick assets: Implicit spread and optimal tick size. *Market Microstructure and Liquidity*, 1(1):1550003, 2015.

[59] J. Donier. Market impact with autocorrelated order flow under perfect competition. *Available at SSRN 2191660*, 2012.

[60] J. Donier, J. Bonart, I. Mastromatteo, and J.-P. Bouchaud. A fully consistent, minimal model for non-linear market impact. *Quantitative Finance*, 15(7):1109–1121, 2015.

[61] R. Donnelly. Ambiguity aversion in algorithmic and high frequency trading. *Available at SSRN 2527808*, 2014.

[62] B. Dupire. Pricing with a smile. *Risk*, 7(1):18–20, 1994.

[63] Z. Eisler, J.-P. Bouchaud, and J. Kockelkoren. The price impact of order book events: Market orders, limit orders and cancellations. *Quantitative Finance*, 12(9):1395–1419, 2012.

[64] A. Esser and B. Mönch. The navigation of an iceberg: The optimal use of hidden orders. *Finance Research Letters*, 4(2):68–81, 2007.

[65] J. D. Farmer, A. Gerig, F. Lillo, and S. Mike. Market efficiency and the long-memory of supply and demand: Is price impact variable and permanent or fixed and temporary? *Quantitative Finance*, 6(2):107–112, 2006.

[66] J. D. Farmer, A. Gerig, F. Lillo, and H. Waelbroeck. How efficiency shapes market impact. *Quantitative Finance*, 13(11):1743–1758, 2013.

[67] J. D. Farmer and F. Lillo. On the origin of power-law tails in price fluctuations. *Quantitative Finance*, 4(1):7–11, 2004.

[68] J. D. Farmer, P. Patelli, and I. Zovko. The predictive power of zero intelligence in financial markets. *Proceedings of the National Academy of Sciences of the United States of America*, 102(6):2254–2259, 2005.

[69] J.-D. Fermanian, O. Guéant, and A. Rachez. Agents behavior on multi-dealer-to-client bond trading platforms. Working paper, 2015.

[70] P. Fodra and H. Pham. High frequency trading in a Markov renewal model. *Available at SSRN 2333752*, 2013.

[71] P. Forsyth, J. Kennedy, S. Tse, and H. Windcliff. Optimal trade execution: A mean quadratic variation approach. *Journal of Economic Dynamics and Control*, 36(12):1971–1991, 2012.

[72] T. Foucault, O. Kadan, and E. Kandel. Liquidity cycles and make/take fees in electronic markets. *The Journal of Finance*, 68(1):299–341, 2013.

[73] C. Frei and N. Westray. Optimal execution of a VWAP order: A stochastic control approach. *Mathematical Finance*, 25(3):612–639, 2015.

[74] J. Gatheral. No-dynamic-arbitrage and market impact. *Quantitative Finance*, 10(7):749–759, 2010.

[75] J. Gatheral and A. Schied. Optimal trade execution under geometric Brownian motion in the Almgren and Chriss framework. *International Journal of Theoretical and Applied Finance*, 14(03):353–368, 2011.

[76] J. Gatheral, A. Schied, and A. Slynko. Transient linear price impact and Fredholm integral equations. *Mathematical Finance*, 22(3):445–474, 2012.

[77] L. Glosten and P. Milgrom. Bid, ask and transaction prices in a specialist market with heterogeneously informed traders. *Journal of Financial Economics*, 14(1):71–100, 1985.

[78] R. Grinold and R. Kahn. *Active portfolio management.* McGraw Hill New York, 2000.

[79] S. Grossman and M. Miller. Liquidity and market structure. *The Journal of Finance*, 43(3):617–633, 1988.

[80] O. Guéant. Permanent market impact can be nonlinear. *arXiv preprint arXiv:1305.0413*, 2013.

[81] O. Guéant. Execution and block trade pricing with optimal constant rate of participation. *Journal of Mathematical Finance*, 4:255–264, 2014.

[82] O. Guéant. Optimal execution and block trade pricing: A general framework. *Applied Mathematical Finance*, 22(4):336–365, 2015.

[83] O. Guéant. Market making on option markets. Working paper, 2016.

[84] O. Guéant. Optimal market making. Working paper, 2016.

[85] O. Guéant, J.-M. Lasry, and J. Pu. A convex duality method for optimal liquidation with participation constraints. *Market Microstructure and Liquidity*, 1(1):1550002, 2015.

[86] O. Guéant and C.-A. Lehalle. General intensity shapes in optimal liquidation. *Mathematical Finance*, 25(3):457–495, 2015.

[87] O. Guéant, C.-A. Lehalle, and J. Fernandez-Tapia. Optimal portfolio liquidation with limit orders. *SIAM Journal on Financial Mathematics*, 3(1):740–764, 2012.

[88] O. Guéant, C.-A. Lehalle, and J. Fernandez-Tapia. Dealing with the inventory risk: A solution to the market making problem. *Mathematics and Financial Economics*, 7(4):477–507, 2013.

[89] O. Guéant and J. Pu. Option pricing and hedging with execution costs and market impact. *To appear in Mathematical Finance*, 2016.

[90] O. Guéant, J. Pu, and G. Royer. Accelerated share repurchase: Pricing and execution strategy. *International Journal of Theoretical and Applied Finance*, 18(3):1550019, 2015.

[91] O. Guéant and G. Royer. VWAP execution and Guaranteed VWAP. *SIAM Journal on Financial Mathematics*, 5(1):445–471, 2014.

[92] F. Guilbaud and H. Pham. Optimal high-frequency trading with limit and market orders. *Quantitative Finance*, 13(1):79–94, 2013.

[93] J. Guyon. Path-dependent volatility. *Risk Technical Paper*, September 2014.

[94] J. Guyon and P. Henry-Labordère. The smile calibration problem solved. *Available at SSRN 1885032*, 2011.

[95] J. Guyon and P. Henry-Labordère. *Nonlinear Option Pricing*. CRC Press, 2013.

[96] P. Hagan, D. Kumar, A. Lesniewski, and D. Woodward. Managing smile risk. *Wilmott*, pages 84–108, September 2002.

[97] L. Harris. *Trading and Exchanges: Market Microstructure for Practitioners*. Oxford University Press, 2002.

[98] L. Harris. *Regulated Exchanges: Dynamic Agents of Economic Growth*. Oxford University Press, 2010.

[99] J. Harrison and D. Kreps. Martingales and arbitrage in multiperiod securities markets. *Journal of Economic Theory*, 20(3):381–408, 1979.

[100] J. Harrison and S. Pliska. Martingales and stochastic integrals in the theory of continuous trading. *Stochastic Processes and Their Applications*, 11(3):215–260, 1981.

[101] D. Heath, R. Jarrow, and A. Morton. Bond pricing and the term structure of interest rates: A discrete time approximation. *Journal of Financial and Quantitative Analysis*, 25(4):419–440, 1990.

[102] D. Heath, R. Jarrow, and A. Morton. Bond pricing and the term structure of interest rates: A new methodology for contingent claims valuation. *Econometrica*, 60(1):77–105, 1992.

[103] P. Henry-Labordère. Calibration of local stochastic volatility models to market smiles: A Monte-Carlo approach. *Risk Magazine*, September 2009.

[104] S. Heston. A closed-form solution for options with stochastic volatility with applications to bond and currency options. *Review of Financial Studies*, 6(2):327–343, 1993.

[105] T. Ho and H. Stoll. Optimal dealer pricing under transactions and return uncertainty. *Journal of Financial Economics*, 9(1):47–73, 1981.

[106] T. Ho and H. Stoll. The dynamics of dealer markets under competition. *The Journal of Finance*, 38(4):1053–1074, 1983.

[107] C. Hopman. Do supply and demand drive stock prices? *Quantitative Finance*, 7(1):37–53, 2007.

[108] U. Horst and F. Naujokat. When to cross the spread? Trading in two-sided limit order books. *SIAM Journal on Financial Mathematics*, 5(1):278–315, 2014.

[109] G. Huberman and W. Stanzl. Price manipulation and quasi-arbitrage. *Econometrica*, 72(4):1247–1275, 2004.

[110] R. Huitema. Optimal portfolio execution using market and limit orders. *Available at SSRN 1977553*, 2014.

[111] H. Hult and J. Kiessling. Algorithmic trading with Markov chains. Working paper, 2010.

[112] M. Humphery-Jenner. Optimal VWAP trading under noisy conditions. *Journal of Banking & Finance*, 35(9):2319–2329, 2011.

[113] S. Jaimungal, D. Kinzebulatov, and D. Rubisov. Optimal accelerated share repurchase. *Available at SSRN 2360394*, 2013.

[114] S. Kakade, M. Kearns, Y. Mansour, and L. Ortiz. Competitive algorithms for VWAP and limit order trading. In *Proceedings of the 5th ACM Conference on Electronic Commerce*, pages 189–198. ACM, 2004.

[115] D. Keim and A. Madhavan. The upstairs market for large-block transactions: Analysis and measurement of price effects. *Review of Financial Studies*, 9(1):1–36, 1996.

[116] F. Klöck, A. Schied, and Y. Sun. Price manipulation in a market impact model with dark pool. *Available at SSRN 1785409*, 2011.

[117] H. Konishi. Optimal slice of a VWAP trade. *Journal of Financial Markets*, 5(2):197–221, 2002.

[118] P. Kovaleva and G. Iori. Optimal trading strategies in a limit order market with imperfect liquidity. Working paper, 2012.

[119] P. Kratz and T. Schöneborn. Optimal liquidation in dark pools. *Quantitative Finance*, 14(9):1519–1539, 2014.

[120] P. Kratz and T. Schöneborn. Portfolio liquidation in dark pools in continuous time. *Mathematical Finance*, 25(3):496–544, 2015.

[121] A. Kyle. Continuous auctions and insider trading. *Econometrica: Journal of the Econometric Society*, 53(6):1315–1335, 1985.

[122] M. Labadie and C.-A. Lehalle. Optimal starting times, stopping times and risk measures for algorithmic trading. *The Journal of Investment Strategies*, 3(2), 2014.

[123] A. Lachapelle, J.-M. Lasry, C.-A. Lehalle, and P.-L. Lions. Efficiency of the price formation process in presence of high frequency participants: A mean field game analysis. *arXiv preprint arXiv:1305.6323*, 2013.

[124] S. Laruelle, C.-A. Lehalle, and G. Pagès. Optimal split of orders across liquidity pools: A stochastic algorithm approach. *SIAM Journal on Financial Mathematics*, 2(1):1042–1076, 2011.

[125] C. Lee and M. Ready. Inferring trade direction from intraday data. *The Journal of Finance*, 46(2):733–746, 1991.

[126] C.-A. Lehalle. The impact of liquidity fragmentation on optimal trading. *Trading*, 2009(1):80–87, 2009.

[127] C.-A. Lehalle. Rigorous strategic trading: Balanced portfolio and mean-reversion. *The Journal of Trading*, 4(3):40–46, 2009.

[128] C.-A. Lehalle and S. Laruelle. *Market Microstructure in Practice*. World Scientific Publishing Ltd., 2013.

[129] C.-A. Lehalle, M. Lasnier, P. Bessson, H. Harti, W. Huang, N. Joseph, and L. Massoulard. What does the saw-tooth pattern on US markets on 19 July 2012 tell us about the price formation process? *Crédit Agricole Cheuvreux Quant Note*, 2012.

[130] H. Leland. Option pricing and replication with transactions costs. *The Journal of Finance*, 40(5):1283–1301, 1985.

[131] D. Li. On default correlation: A copula function approach. *Available at SSRN 187289*, 1999.

[132] T. Li. *Dynamic Programming and Trade Execution*. PhD thesis, Princeton University, 2013.

[133] T. Li. Optimal limit-versus-market order slicing under a VWAP benchmark-discrete case. *Available at SSRN 2318890*, 2013.

[134] T. Li and R. Almgren. Option hedging with smooth market impact. Working paper, 2015.

[135] F. Lillo and J. D. Farmer. The long memory of the efficient market. *Studies in Nonlinear Dynamics & Econometrics*, 8(3), 2004.

[136] F. Lillo, J. D. Farmer, and R. Mantegna. Single curve collapse of the price impact function for the New York Stock Exchange. *arXiv preprint cond-mat/0207428*, 2002.

[137] F. Lillo, J. D. Farmer, and R. Mantegna. Econophysics: Master curve for price-impact function. *Nature*, 421(6919):129–130, 2003.

[138] P.-L. Lions and J.-M. Lasry. Large investor trading impacts on volatility. *Annales de l'IHP Analyse non linéaire*, 24(2):311–323, 2007.

[139] A. Lipton. The vol smile problem. *Risk*, 15(2):61–66, 2002.

[140] J. Lorenz and R. Almgren. Mean-variance optimal adaptive execution. *Applied Mathematical Finance*, 18(5):395–422, 2011.

[141] J. McCulloch. Relative volume as a doubly stochastic binomial point process. *Quantitative Finance*, 7(1):55–62, 2007.

[142] J. McCulloch and V. Kazakov. Optimal VWAP trading strategy and relative volume. Working paper, 2007.

[143] J. McCulloch and V. Kazakov. Mean variance optimal VWAP trading. *Available at SSRN 1803858*, 2012.

[144] McKinsey & Company. Corporate bond e-trading: Same game, new playing field. Technical report, 2013.

[145] A. Menkveld. High frequency trading and the new market makers. *Journal of Financial Markets*, 16(4):712–740, 2013.

[146] D. Mitchell, J. Bialkowski, and S. Tompaidis. Optimal VWAP tracking. *Available at SSRN 2333916*, 2013.

[147] O. Morgenstern and J. von Neumann. *Theory of Games and Economic Behavior*. Princeton University Press, 1953.

[148] E. Moro, J. Vicente, L. Moyano, A. Gerig, J. D. Farmer, G. Vaglica, F. Lillo, and R. Mantegna. Market impact and trading profile of hidden orders in stock markets. *Physical Review E*, 80(6), 2009.

[149] Y. Nevmyvaka, Y. Feng, and M. Kearns. Reinforcement learning for optimized trade execution. In *Proceedings of the 23rd International Conference on Machine Learning*, pages 673–680. ACM, 2006.

[150] K. Nyström, S. M. Ould Aly, and C. Zhang. Market making and portfolio liquidation under uncertainty. *International Journal of Theoretical and Applied Finance*, 17(5):1450034, 2014.

[151] A. Obizhaeva and J. Wang. Optimal trading strategy and supply/demand dynamics. *Journal of Financial Markets*, 16(1):1–32, 2013.

[152] E. Platen and M. Schweizer. On feedback effects from hedging derivatives. *Mathematical Finance*, 8(1):67–84, 1998.

[153] V. Plerou, P. Gopikrishnan, X. Gabaix, and H. Stanley. Quantifying stock-price response to demand fluctuations. *Physical Review E*, 66:027104, 2002.

[154] M. Potters and J.-P. Bouchaud. More statistical properties of order books and price impact. *Physica A: Statistical Mechanics and Its Applications*, 324(1):133–140, 2003.

[155] R. Rockafellar. Conjugate convex functions in optimal control and the calculus of variations. *Journal of Mathematical Analysis and Applications*, 32(1):174–222, 1970.

[156] R. Rockafellar. *Convex Analysis*. Princeton University Press, 1970.

[157] R. Rockafellar. Existence and duality theorems for convex problems of Bolza. *Transactions of the American Mathematical Society*, 159:1–40, 1971.

[158] L. Rogers and S. Singh. The cost of illiquidity and its effects on hedging. *Mathematical Finance*, 20(4):597–615, 2010.

[159] S. Ronnie and G. Papanicolaou. General Black-Scholes models accounting for increased market volatility from hedging strategies. *Applied Mathematical Finance*, 5(1):45–82, 1998.

[160] W. Schachermayer and J. Teichmann. How close are the option pricing formulas of Bachelier and Black-Merton-Scholes? *Mathematical Finance*, 18(1):155–170, 2008.

[161] A. Schied and T. Schöneborn. Risk aversion and the dynamics of optimal liquidation strategies in illiquid markets. *Finance and Stochastics*, 13(2):181–204, 2009.

[162] A. Schied, T. Schöneborn, and M. Tehranchi. Optimal basket liquidation for CARA investors is deterministic. *Applied Mathematical Finance*, 17(6):471–489, 2010.

[163] C. Sprenkle. Warrant prices as indicators of expectations and preferences. *Yale Economic Essays*, 1(2):178–231, 1961.

[164] S. Stoikov and M. Sağlam. Option market making under inventory risk. *Review of Derivatives Research*, 12(1):55–79, 2009.

[165] H. Stoll. Inferring the components of the bid-ask spread: Theory and empirical tests. *The Journal of Finance*, 44(1):115–134, 1989.

[166] R. Tachet des Combes. *Non-Parametric Model Calibration in Finance.* PhD thesis, École Centrale, 2011.

[167] B. Tóth, Y. Lemperiere, C. Deremble, J. de Lataillade, J. Kockelkoren, and J.-P. Bouchaud. Anomalous price impact and the critical nature of liquidity in financial markets. *Physical Review X*, 1(2), 2011.

[168] B. Tóth, I. Palit, F. Lillo, and J. D. Farmer. Why is equity order flow so persistent? *Journal of Economic Dynamics and Control*, 51:218–239, 2015.

[169] S. Tse, P. Forsyth, J. Kennedy, and H. Windcliff. Comparison between the mean-variance optimal and the mean-quadratic-variation optimal trading strategies. *Applied Mathematical Finance*, 20(5):415–449, 2013.

[170] O. Vasicek. An equilibrium characterization of the term structure. *Journal of Financial Economics*, 5(2):177–188, 1977.

[171] J. Wald and H. Horrigan. Optimal limit order choice. *The Journal of Business*, 78(2):597–620, 2005.

[172] P. Weber and B. Rosenow. Order book approach to price impact. *Quantitative Finance*, 5(4):357–364, 2005.

[173] P. Wilmott. *Frequently Asked Questions in Quantitative Finance.* John Wiley & Sons, 2010.

[174] P. Wilmott and P. Schönbucher. The feedback effect of hedging in illiquid markets. *SIAM Journal on Applied Mathematics*, 61(1):232–272, 2000.

[175] A. Wranik. A trading system for flexible VWAP executions as a design artifact. *PACIS 2009 Proceedings*, 2009.

[176] C. Yingsaeree. *Algorithmic Trading: Model of Execution Probability and Order Placement Strategy.* PhD thesis, University College London, 2012.

[177] E. Zarinelli, M. Treccani, J. D. Farmer, and F. Lillo. Beyond the square root: Evidence for logarithmic dependence of market impact on size and participation rate. *arXiv preprint arXiv:1412.2152*, 2014.

Index

Milton Keynes UK
Ingram Content Group UK Ltd.
UKHW031129141024
449569UK00006B/344

9 781498 725477